ATLAS AND TITAN SPACE OPERATIONS AT CAPE CANAVERAL 1993 — 2006

by Mark C. Cleary

45th SPACE WING
History Office

ATLAS AND TITAN SPACE OPERATIONS AT THE CAPE, 1993 – 2006

TABLE OF CONTENTS

PREFACE.. iv

ATLAS I/CENTAUR and ATLAS II/CENTAUR Overview.. 1

ATLAS I/CENTAUR and ATLAS II/CENTAUR Military Space Operations............... 3
UHF F/O #1; DSCS III (Jul 93); UHF F/O #2; DSCS III (Nov 93); UHF F/O #3; EHF F/O #4; EHF F/O #5; DSCS III (Jul 95); EHF F/O #6; UHF F/O #7; DSCS III (Oct 97); NRO Payload (Jan 98); UHF F-8; UHF F-9; UHF F-10; DSCS III B8 (Jan 00); DSCS III B11 (Oct 00); NRO MLV-11 (Dec 00); NRO MLV-12 (Oct 01); NROL-1.

ATLAS I/CENTAUR and ATLAS II/CENTAUR Commercial Space Operations.......... 18
TELSTAR 4; DIRECTV; INTELSAT VII (Oct 94); ORION; INTELSAT VII (Jan 95); INTELSAT VII (Mar 95); MSAT; JCSAT; GALAXY IIIR; PALAPA-C1; INMARSAT-3; SAX; GE-1; HOT BIRD 2; INMARSAT-3 F3; JCSAT-4; TEMPO II; SUPERBIRD-C; GE-3; ECHOSTAR III; GALAXY VIII-i; INTELSAT 806; INTELSAT 805; HOTBIRD-5; JCSAT-6; EUTELSAT W3; ECHOSTAR V; HISPASAT 1-C; ECHOSTAR VI; ICO-A1 (F2); HISPASAT-1D; AMC-10; SUPERBIRD-6; AMC-11.

ATLAS I/CENTAUR and ATLAS II/CENTAUR Civil Space Operations..................... 41
GOES-I; GOES-J; SOHO; GOES-K; GOES-L; TDRS-H; GOES-M; TDRS-I; TDRS-J.

ATLAS IIIA/CENTAUR and ATLAS IIIB/CENTAUR Overview............................... 48

ATLAS IIIA and ATLAS IIIB/CENTAUR Commercial Space Operations................... 50
EUTELSAT W4; ECHOSTAR VII; ASIASAT-4; MBSat.

ATLAS IIIB/CENTAUR Military Space Operations... 53
UHF F-11; NROL-23.

ATLAS V/CENTAUR Overview... 56
The Lockheed Martin Evolved Expendable Launch Vehicle (EELV) Program
Deactivation of Complex 41 and Construction for ATLAS V
The ATLAS V/CENTAUR Family of Launch Vehicles

ATLAS V Commercial Space Operations... 62
HOT BIRD-6; HELLAS-SAT; RAINBOW-1; AMC-16; INMARSAT-4; ASTRA 1KR.

ATLAS V Civil Space Operations.. 68
MARS RECON ORBITER; PLUTO NEW HORIZONS.

TITAN IV/CENTAUR and TITAN IV/IUS Overview................................. 71

TITAN IV Military Space Operations... 73
MILSTAR (Flt 1); Classified Payload (May 94); Classified Payload (Aug 94); DSP (Dec 94); Classified Payload (May 95); Classified Payload (Jul 95); MILSTAR (Flt 2); Classified Payload (Apr 96); Classified Payload (Jul 96); DSP (Feb 97); Classified Payload (Nov 97); NRO Payload (May 98); NRO Payload (Aug 98); DSP-19; MILSTAR (Flt 3); DSP-20; MILSTAR (Flt 4); DSP-21; MILSTAR (Flt 5); MILSTAR (Flt 6); NROL-19; DSP-22; NROL-16.

TITAN IV Civil Space Operations... 102
CASSINI.

ENDNOTES.. 104

ATLAS AND TITAN Launch Synopsis.. 137

PREFACE

This study addresses ATLAS and TITAN programs, ATLAS V facility improvements, and individual missions involving ATLAS and TITAN space launches at Cape Canaveral from 1993 through 2006. I reedited this information from annual histories I authored for the years under consideration. Competent authorities have reviewed all the material presented in this study, and it is releasable to the general public. Any factual errors noted by the reader may be addressed to the 45th Space Wing History Office, 1201 Edward H. White II Street, PAFB, FL 32925-3299.

The section entitled, "The Lockheed Martin Evolved Expendable Launch Vehicle (EELV) Program," provides a brief overview of government and contractor efforts that led to ATLAS V operations on Complex 41 at the Cape. Other sections address the deactivation and demolition of Complex 36, ATLAS V construction on Complex 41, descriptions of the various types of launch vehicles, a chronological list of 101 major launches, and individual mission summaries.

The mission summaries are grouped by type of vehicle, and they are listed chronologically under three subjective headings: 1) Military Space Operations, 2) Commercial Space Operations, and 3) Civil Space Operations. Since NASA, the Air Force, and the Navy use commercial launch contractors to put their payloads into space, there has been some confusion over just what constitutes a civil, military or commercial launch operation. For our purposes, the sponsor of the mission (Air Force, Navy, NASA or commercial company) will determine where its launch summary is placed.

Each launch entry contains a brief synopsis of the mission. Highlights of the preparations for each flight and the launch itself, including delays, scrubs, incidents, and the outcome of the flight, are included. The ATLAS and TITAN Launch Synopsis at the end of this study lists all the launches in chronological order by date (calculated in Greenwich Mean Time – indicated by a "Z" suffix at the end of each time entry), type of launch vehicle, launch site, and payload. Each launch summary and topic is hyperlinked in the Table of Contents and Launch Synopsis for ease of retrieval.

MARK C. CLEARY
September 2007

ATLAS I/CENTAUR and ATLAS II/CENTAUR Overview

Though the first ATLAS II/CENTAUR was launched from Cape Canaveral on 7 December 1991, it must be emphasized that the ATLAS II/CENTAUR and the ATLAS I/CENTAUR were both based on the old ATLAS Intercontinental Ballistic Missile (ICBM) of the late 1950s. In particular, the first stage of the later vehicles shared the ATLAS ICBM's pressurized booster/sustainer design. Since the stainless steel walls of the booster/sustainer were very thin, internal tank pressure or mechanical stretching was used to maintain the vehicle's structural integrity on the ground. An onboard pressurization system sustained the vehicle's structural integrity during flight. Both sections of the booster/sustainer engine systems used on the ATLAS I and ATLAS II burned RP-1 (i.e., highly refined kerosene) for fuel. The booster/sustainer on the ATLAS I/CENTAUR was 73 feet long, and its MA-5 engine system provided 377,500 pounds of thrust (booster) and 60,000 pounds of thrust (sustainer) respectively. The booster/sustainer on the ATLAS II/CENTAUR was 81.7 feet long, and its MA-5A engine system provided 473,000 pounds of thrust at sea level.

Both ATLAS launch vehicles used identical Pratt & Whitney RL10A-3-3A engines in their CENTAUR upper stages. One significant difference was that the ATLAS II's upper stage was 3 feet longer than the CENTAUR on the ATLAS I (e.g., 33 feet versus 30 feet). In either configuration, each CENTAUR had two Pratt & Whitney RL10A-3-3A engines to provide a total of 33,000 pounds of thrust at altitude. Liquid hydrogen fueled the CENTAUR's main engines, and hydrazine fueled the CENTAUR's Reaction Control System. Both vehicles could use either an 11-foot or a 14-foot payload fairing. Depending on the size of its fairing, the ATLAS I/CENTAUR could place between 4,970 pounds and 5,240 pounds into geosynchronous transfer orbit. The ATLAS II/CENTAUR could place between 6,200 pounds and 6,500 pounds into geosynchronous transfer orbit. An improved version of the vehicle — the ATLAS IIA/CENTAUR — was equipped with two Pratt & Whitney RL10A-4 engines rated at 20,800 pounds of thrust each. An enhanced performance version — the ATLAS IIAS/CENTAUR — was equipped with two RL10A-4s and four Thiokol Castor IVA solid rocket motors to increase the vehicle's maximum payload lift capability from 6,970 pounds to a substantial 8,450 pounds to geosynchronous transfer orbit.[1] Though their payload capabilities differed, all three versions of the ATLAS II/CENTAUR stood 156 feet tall on the launch pad when equipped with the large fairing. The ATLAS I/CENTAUR stood 144 feet tall. The first ATLAS IIA/CENTAUR was launched from the Cape on 10 June 1992. The first ATLAS IIAS/CENTAUR mission was launched from Pad 36B on 16 December 1993.[2]

Though the precise timing of ATLAS I/CENTAUR and ATLAS II/CENTAUR staging events varied from mission to mission, certain sequences were common to all ATLAS/CENTAUR flights. Following lift-off, the booster and sustainer engines provided combined thrust to send the vehicle on its way. Between two and one-half minutes and three minutes into the flight, the section containing the two booster engines was jettisoned, but the sustainer engine continued to thrust for approximately two minutes until the stage's fuel was depleted. The payload fairing was jettisoned during the sustainer burn. Two or three seconds after sustainer engine cut-off, the CENTAUR separated from the ATLAS. Depending on the mission, the CENTAUR's first main engine burn started about 16 seconds later. The first CENTAUR burn lasted about four and one-half minutes. The first burn placed the CENTAUR and its payload into an elliptical low-Earth parking orbit. The CENTAUR's axial thrusters were brought into play for propellant settling, venting, and pre-start purposes during a coasting period that typically ranged from about 11 minutes to 15 minutes. The CENTAUR's second main engine burn started typically between 22 to 25 minutes into the flight, and the CENTAUR and its payload were accelerated into a highly elliptical transfer orbit over the next 90 seconds of the flight. On *most* ATLAS/CENTAUR missions, the CENTAUR's second burn continued until the CENTAUR nearly ran out of fuel. This process was known as "minimum residual shutdown". Following main engine cut-off (about 26 minutes after lift-off), the CENTAUR separated from the spacecraft and performed a Collision and Contamination Avoidance Maneuver (CCAM) to carry itself out of the spacecraft's orbit. The CCAM included a tank 'blowdown' action, which vented the CENTAUR's tanks of any remaining pressure and fuel. The blowdown eliminated the possibility of tank fragmentation, which might otherwise leave debris in the spacecraft's flight path.[3]

Range instrumentation for a typical ATLAS II/CENTAUR mission included radars on the Cape, Merritt Island, Patrick AFB, and Antigua. Jonathan Dickinson also provided radar support for some ATLAS II/CENTAUR missions. For northerly flights, radar assets on Wallops Island and Station 53 located in Argentia, Newfoundland provided additional coverage. Command transmitter/exciters at the Cape and Jonathan Dickinson provided command and control for all missions. Depending of the flight azimuth, additional command support came from Antigua or Wallops Island. HH-60G helicopters operated out of Patrick AFB to provide safety sea surveillance for all ATLAS II/CENTAUR missions, and an SCE Learjet provided weather reconnaissance. Telemetry units on Merritt Island, Jonathan Dickinson, and Antigua supported ATLAS II/CENTAUR missions. The Tracking and Data Relay Satellite System (TDRSS) and telemetry assets on Ascension and Wallops Island also covered some of the flights. Optical support included Patrick's Fixed Intercept Ground Optical Recorder (IGOR), the Cocoa Beach or Playalinda Beach Distant Object Attitude Measuring System (DOAMS), and cameras at various range optical sites. Patrick's WSR-74C weather radar, Melbourne's WSR-88D radar, and a whole host of weather instrumentation systems at Cape Canaveral (e.g., Doppler Radar Wind Profiler, Launch Pad Lightning Warning System, Meteorological Interactive Date Display System, the Lightning Detection and Range System, weather trackers, and computers) supported each flight.[4]

ATLAS I/CENTAUR and ATLAS II/CENTAUR Military Space Operations

ATLAS I/CENTAUR (UHF F/O #1), AC-74, 25 March 1993

The object of the AC-74 mission was to boost a U.S. Navy Ultra High Frequency (UHF) communications satellite into a highly elliptical transfer orbit to upgrade the Defense Department's existing constellation of Fleet Satellite Communications (FLTSATCOM) and leased satellite spacecraft.[5] An ATLAS I/CENTAUR commercial launch vehicle was chosen for the task, and it was prepared for launch from Pad 36B. Technicians erected the Atlas I booster/sustainer on 19 October 1992, and technicians and engineers raised the vehicle's interstage and CENTAUR upper stage on 20 and 21 October respectively. Officials conducted a simulated flight test on 24 November, and they completed the mission's first Wet Dress Rehearsal on 1 December 1992. The second simulated flight test was finished successfully on 10 February 1993, and the second Wet Dress Rehearsal was completed on 16 February 1993. After technicians encapsulated the spacecraft at the end of February, engineers mated the payload to the vehicle on 2 March 1993. Technicians completed hydrazine propellant loading operations on 11 March 1993, but the launch was delayed about 11 days while engineers completed Pratt-Whitney engine testing. Due to range scheduling conflicts with the STS-55 Shuttle mission, General Dynamics agreed to delay the UHF Follow-On launch until 25 March 1993.[6]

Thick clouds at 3000 feet made the weather marginal on 25 March, but the weather cleared sufficiently to permit a safe launch. Following the countdown, the ATLAS I/CENTAUR lifted off Pad 36B at 2138:01.950Z on 25 March 1993. Though the initial lift-off appeared normal, telemetry and safety displays soon revealed that the vehicle's booster chamber pressure was only 75 percent or less of what was expected. The ATLAS I could not follow a normal flight path under reduced power, but the mission rules were "never violated" during the booster/ sustainer phases of the flight. To compensate for the ATLAS' anemic performance, the CENTAUR upper stage's first burn lasted longer than anticipated, and it put the spacecraft in the prescribed parking orbit. Unfortunately, the CENTAUR's first burn left it short on fuel, and the second burn did not achieve the planned transfer orbit. The Navy subsequently declared the mission a total failure.[7]

ATLAS II/CENTAUR (DSCS III), AC-104, 19 July 1993

The object of the AC-104 mission was to place a Defense Satellite Communications System (DSCS) III spacecraft [8] into a highly elliptical transfer orbit. An ATLAS II/CENTAUR vehicle was chosen for the mission, and it would be launched from Pad 36A. General Dynamics erected the ATLAS II booster/sustainer on 31 July 1992, and engineers raised the CENTAUR upper stage on 3 August 1992. Unfortunately, the AC-71 mission failure on 22 August 1992 delayed the AC-104 mission. Following a launch vehicle simulated test on 4 September 1992, tanking tests and Mobile Service Tower repairs occupied most of the next three months. Officials completed the vehicle's first Wet Dress Rehearsal on 12 January 1993, but a second ATLAS I/CENTAUR failure on 25 March 1993 delayed the AC-104 mission once again. The second Wet Dress Rehearsal for AC-104 was finished on 25 May 1993. Technicians encapsulated the spacecraft during the second week in July, and engineers mated the payload to the vehicle shortly thereafter. Technicians completed hydrazine loading operations around 15 July, and the countdown got underway on 19 July 1993. One control console experienced a minor display problem, but all range instrumentation systems were ready to support the launch. The ATLAS II/CENTAUR lifted off Pad 36A at 2204:02.162Z on 19 July 1993. The flight was successful, and the DSCS III payload completed the initial DSCS III constellation of communications satellites. The mission was also significant as the first successful "return to flight" for the CENTAUR upper stage after the program "stood down" due to ATLAS/CENTAUR launch failures on 18 April 1991 and 22 August 1992.[9]

ATLAS I/CENTAUR (UHF F/O #2), AC-75, 3 September 1993

The object of the AC-75 mission was to boost a UHF Follow-On communications satellite into a highly elliptical transfer orbit to upgrade the existing constellation of FLTSATCOM and leased satellite spacecraft that provided global communications for U.S. Navy land-based and mobile users. A commercial ATLAS I/CENTAUR was chosen for the mission, and it was prepared for launch at Pad 36B. General Dynamics erected the vehicle's ATLAS booster/sustainer on 7 and 8 June 1993. Technicians and engineers added the interstage and CENTAUR upper stage to the vehicle on 9 and 10 June respectively. Officials completed a launch vehicle simulated flight test on 29 July, and they finished the Wet Dress Rehearsal on 7 August 1993. Technicians encapsulated the spacecraft, and engineers mated the payload to the launch vehicle on 24 August 1993. Officials wrapped up the composite electrical readiness test on 26 August 1993. Technicians completed hydrazine loading operations on 29 August, and officials readied the vehicle for launch on 3 September 1993.[10]

The countdown proceeded on 3 September, but the count was stopped at T minus 31 seconds after the ATLAS' second stage readiness light did not illuminate. Fortunately, other indicators confirmed the second stage was ready, and the anomaly was chalked up to a faulty indicator, not a faulty booster. Following an unplanned hold of 14 minutes, the count was recycled to T minus five minutes. The countdown resumed at 1112Z, and the ATLAS I/CENTAUR was launched at 1117:02.033Z on 3 September 1993. The launch was successful.[11]

ATLAS II/CENTAUR (DSCS III), AC 106, 28 November 1993

The object of the AC-106 mission was to boost a DSCS III communications satellite into a highly elliptical transfer orbit. An ATLAS II/CENTAUR vehicle was chosen for the task, and it was prepared for launch on Pad 36A. General Dynamics erected the ATLAS booster/sustainer on 29 September 1993. The company's technicians and engineers erected the interstage and CENTAUR upper stage on 30 September and 1 October 1993 respectively. Officials conducted a launch vehicle simulated flight test on 3 November, and they completed the Wet Dress Rehearsal for the mission on 10 November 1993. Following spacecraft encapsulation on 12 and 13 November, engineers mated the DSCS III payload to the launch vehicle on 15 and 16 November 1993. Technicians completed hydrazine loading operations on 22 November, and the vehicle was readied for launch on 28 November 1993.[12]

The countdown on the 28th went well. Nevertheless, officials were forced to call a 26-minute-long unscheduled hold (at T minus 40 seconds) for a CENTAUR GHe check valve failure indication. The "failure" was the result of faulty data, and the problem was corrected quickly. Officials recycled the count to T minus five minutes, and the countdown resumed without further ado. The ATLAS II/CENTAUR lifted off Pad 36A at 2340:01.971Z on 28 November 1993. All assigned radars met their commitments in a highly satisfactory manner, and the mission was a success. The DSCS III spacecraft replaced an older DSCS III that had exceeded its designed life expectancy.[13]

ATLAS I/CENTAUR (UHF F/O #3), AC-76, 24 June 1994

The object of the AC-76 mission was to place the U.S. Navy's UHF Follow On #3 spacecraft into a highly elliptical transfer orbit. Following spacecraft/vehicle separation, four firing sequences commonly called 'burns' would boost the spacecraft to an apogee of 19,322 nautical miles. Three more burns at the vehicle's orbital apogee would raise the spacecraft's perigee and circularize the orbit. The UHF Follow On program was designed to replace and upgrade the existing constellation of Fleet Satellite Communications (FLTSATCOM) and Leased Satellite

(LEASAT) spacecraft, which provided global communications for Navy vessels and shore-based installations. The satellite was the third of ten communications spacecraft built under a $1.7 billion contract with Hughes Space Communications. The fueled UHF F/O #3 spacecraft weighed approximately 6,300 pounds, and it had an operational life expectancy of 14 years. With its solar panels deployed, the satellite had a wingspan of 60 feet. Previously, General Dynamics launched all ATLAS missiles and space vehicles, but Martin Marietta purchased General Dynamics Space Systems Division on 2 May 1994. Consequently, the launch was the first ATLAS/CENTAUR mission completed under Martin Marietta Space Systems.[14]

Technicians erected the ATLAS I booster/sustainer and interstage adapter on Pad 36B on 22 April 1994. Engineers raised the CENTAUR upper stage on 23 April, and officials conducted the Wet Dress Rehearsal toward the end of May 1994. Following encapsulation of the spacecraft and vehicle/payload mating in mid-June, engineers and technicians prepared the vehicle for launch. Though the countdown on 24 June 1994 was delayed five minutes due to a fouled range precaution, the countdown resumed and concluded successfully. The ATLAS I/CENTAUR lifted off Pad 36B at 1350:02.578Z on 24 June 1994. Martin Marietta Space Systems Vice President Michael W. Wynne was proud to report a "100 percent mission success."[15]

ATLAS II/CENTAUR (EHF F/O #4), AC-112, 29 January 1995

The object of the AC-112 mission was to place a U.S. Navy-sponsored communications satellite — the EHF Follow-on 4 (EHF F/O #4) — into an intermediate transfer orbit. The mission was part of a program to replace and upgrade the existing constellation of aging Fleet Satellite Communications (FLTSATCOM) spacecraft and leased satellites that provided global communications for the Navy's land-based and mobile users. The EHF F/O #4 was the fourth of ten communications satellites built under a $1.7 billion contract with Hughes Space Communications. The 6,664-pound spacecraft weighed about 350 pounds more than its three predecessors because it included an Extremely High Frequency (EHF) package. The satellite had an operational life expectancy of 14 years.[16]

The ATLAS II/CENTAUR vehicle arrived at Cape Canaveral on 6 December 1994. Following the ATLAS and interstage erection on 8 December, engineers erected the CENTAUR upper stage on 9 December 1994. Officials accomplished the launch vehicle simulated flight test on 4 January 1995, and they completed the Wet Dress Rehearsal on 15 January 1995. Following the payload's encapsulation on 17 January, engineers mated the spacecraft to the launch vehicle on 18 January 1995. Officials completed the vehicle's launch certification on 20 January 1995. Preparations continued for the launch, which was scheduled for 28 January 1995. The countdown on the 28th required one unscheduled hold of 64 minutes to troubleshoot a bad communications cable and bring Range Safety's MODCOMP B computer back online. Officials completed the countdown, and the ATLAS II/CENTAUR lifted off Pad 36A at 0125:00.581Z on 29 January 1995. The launch was successful.[17]

ATLAS II/CENTAUR (EHF F/O #5), AC-116, 31 May 1995

The object of the AC-116 mission was to place a 6,664-pound communications satellite — the EHF Follow-on 5 (EHF F/O #5) — into an intermediate transfer orbit. The launch was part of a program to replace and upgrade the existing constellation of aging Fleet Satellite Communications (FLTSATCOM) spacecraft and leased satellites that provided global communications for the Navy's land-based and mobile users. The EHF F/O #5 was the fifth of ten communications satellites built under a $1.7 billion contract with Hughes Space Communications. The satellite had an operational life expectancy of 14 years.[18]

The ATLAS II portion of the vehicle arrived at Cape Canaveral on 14 April 1995. Technicians erected the ATLAS and its interstage adapter on 19 and 20 April 1995 respectively. The CENTAUR upper stage arrived on 21 April, and engineers erected it on 22 April 1995. Following the launch vehicle simulated flight test on 8 May, officials completed the Wet Dress Rehearsal on 13 May 1995. Technicians encapsulated the spacecraft on 18 May, and engineers mated the payload to the launch vehicle on 20 May 1995. Officials completed launch vehicle certification on 23 May, and preparations continued for a launch on 31 May 1995. There were no unscheduled holds during the countdown, and the ATLAS II/CENTAUR lifted off Pad 36A at 1527:01.485Z on 31 May 1995. The launch was successful.[19]

ATLAS IIA/CENTAUR (DSCS III), AC-118, 31 July 1995

The object of the AC-118 mission was to place the fifth in a series of eight Defense Satellite Communications System (DSCS) III spacecraft into geosynchronous transfer orbit to upgrade/replace the Air Force's DSCS constellation. The DSCS satellites served all branches of the Defense Department. The third-generation DSCS III payload was designed to provide uninterrupted secure voice and high data rate communications via six independent SHF transponder channels. Lockheed Martin Aerospace built the spacecraft. The DSCS III weighed 2,615 kilograms, and it had an operational life expectancy of 10 years.[20]

The payload arrived at Cape Canaveral's DSCS Processing Facility (DPF) on 8 April 1995. The ATLAS IIA/CENTAUR chosen for the mission was off-loaded from a C-5 transport aircraft on 7 June 1995. Technicians erected the ATLAS IIA and its interstage adapter on Pad 36A on 8 June, and engineers raised the CENTAUR upper stage on 9 June 1995. Following the launch vehicle simulated flight test on 30 June, officials completed the Wet Dress Rehearsal on 8 July 1995. The DSCS III was encapsulated on 11 and 12 July, and engineers mated the payload to the launch vehicle on the 13 July 1995. Following launch vehicle certification on 17 July 17, preparations continued for the first launch attempt on 26 July 1995. Unfortunately, the weather did not cooperate, and officials were forced to scrub the launch at 1955Z on 26 July for high winds and thunderstorms in the local area. The second launch attempt on 27 July was scrubbed at 2215Z due to a high-pressure leak found on a ground storage tank regulator. In addition to that condition, the range was 'Red' due to heavy clouds when the scrub was called. Officials called the third scrub at 2117Z on 28 July for a violation of wind constraints. Finally, after one five-minute unscheduled weather hold and one faulty instrument indication, the ATLAS IIA/CENTAUR lifted off Pad 36A at 2330:01.850Z on 31 July 1995. The launch was successful.[21]

ATLAS II/CENTAUR (EHF F/O #6), AC-119, 22 October 1995

The object of the AC-119 mission was to place a 6,664-pound communications satellite — the EHF Follow-on 6 (EHF F/O #6) — into an intermediate transfer orbit. The launch was part of a program to replace and upgrade the existing constellation of aging Fleet Satellite Communications (FLTSATCOM) spacecraft and leased satellites that provided global communications for the Navy's land-based and mobile users. The EHF F/O #6 was the sixth of ten communications satellites built under a $1.7 billion contract with Hughes Space Communications. The satellite had an operational life expectancy of 14 years.[22]

The ATLAS II/CENTAUR selected for the mission was received at Cape Canaveral on 29 August 1995. Technicians erected the ATLAS II on Pad 36A on 31 August, and they installed the interstage adapter on 1 September 1995. Engineers raised the CENTAUR upper stage on 5 September, and officials accomplished the launch vehicle simulated flight test on 21 September 1995. Following the Wet Dress Rehearsal on 27 September, engineers encapsulated the spacecraft and erected it on 6 October 1995. Launch certification was completed on 8 October, and preparations continued for the first launch attempt on 17 October 1995. The countdown on 17 October went smoothly until 0515Z when high winds caused an unplanned hold and prevented the Mobile Service Tower's rollback. The winds eventually forced officials to scrub the launch at 0703Z on the 17th. As high winds continued through the night and morning, officials called another launch scrub on 18 October due to high winds and thick clouds. Fortunately, the weather cooperated during the third countdown on 22 October. The MST was rolled back, and the ATLAS II/CENTAUR lifted off Pad 36A at 0800:02.043Z on 22 October 1995. The launch was successful.[23]

ATLAS II/CENTAUR (UHF F/O #7), AC-125, 25 July 1996

The object of the AC-125 mission was to place a U.S. Navy UHF Follow-on spacecraft (UHF F/O #7) in an intermediate transfer orbit from which the spacecraft could emerge to boost itself into a geosynchronous transfer orbit and (finally) a circular geosynchronous orbit. The spacecraft was one of a constellation of six communications satellites (designated F4 through F9) introduced to replace and upgrade an aging constellation of Fleet Satellite Communications (FLTSATCOM) and leased satellites. It was launched into space to provide continued global communications for shore-based and mobile Navy units. Hughes' Space and Communications Group built the spacecraft in El Segundo, California. The satellite had an operational life expectancy of 14 years. Astrotech Space Operations processed the payload in Titusville prior to its arrival at the Cape. Lockheed Martin handled payload/launch vehicle integration and the launch from Pad 36A.[24]

The ATLAS II/CENTAUR vehicle arrived at Cape Canaveral on 10 May 1996. Technicians erected the ATLAS and its interstage adapter on 14 and 15 May respectively, and engineers added the CENTAUR upper stage to the launch vehicle on 16 May 1996. Officials completed the Wet Dress Rehearsal on 19 June, and engineers mated the UHF F/O #7 spacecraft to the launch vehicle on 12 July 1996. Following launch certification and RP-1 propellant tanking operations, preparations continued for the launch on 25 July 1996. During the countdown, there was one unscheduled 15-minute hold to resolve an indication that the Stage I vibration suppression system was not fully charged. The anomaly team determined there was no risk to the vehicle, and the countdown continued. The vehicle lifted off Pad 36A at 1241:01.000Z on 25 July 1996. The launch was highly successful. The vehicle actually placed the payload into an orbit 3,000 miles higher than expected, thereby reducing onboard fuel expenditures to raise the spacecraft to final orbit. Theoretically, this extended the spacecraft' operational life since more fuel was available for on-orbit station keeping. The 6,655-pound spacecraft underwent a standard month-long series of on-orbit tests to qualify it for its position in the UHF Follow-on constellation.[25]

ATLAS IIA/CENTAUR (DSCS III), AC-131, 25 October 1997

The object of the AC-131 mission was to place a Defense Satellite Communications System (DSCS) III spacecraft in a transfer orbit from Pad 36A. Ultimately, the DSCS III would replace an older satellite and upgrade the existing DSCS III constellation. Lockheed Martin Missiles and

Space built the DSCS III, and it was processed at the DSCS Processing Facility (DPF) at Cape Canaveral.[26] The satellite was designed to operate from geosynchronous orbit. It provided secure voice and high data rate military communications for a wide variety of users ranging from airborne terminals to large fixed installations. The DSCS constellation served all branches of the Defense Department, diplomatic missions, the White House, NATO, and the United Kingdom.[27]

The ATLAS IIA/CENTAUR vehicle arrived at Cape Canaveral on 4 September 1997. Technicians erected the ATLAS and its interstage adapter on 15 September, and they raised the CENTAUR upper stage on 17 September 1997. After the Wet Dress Rehearsal on 10 October, engineers mated the spacecraft to the launch vehicle on 14 October. Officials certified the vehicle for launch on 16 October 1997. Following RP-1 propellant tanking operations, preparations continued for the launch on 25 October 1997. There was one unscheduled 69-minute hold for thick clouds and liquid oxygen topping valve problem during the countdown on 24/25 October. Both complications were resolved, and the count resumed at 0041Z on the 25th. The vehicle lifted off Pad 36A at 0046:00.282Z on 25 October 1997. The launch was successful.[28]

ATLAS IIA/CENTAUR (NRO Payload), AC-109, 29 January 1998

The object of the AC-109 mission was to place a National Reconnaissance Office (NRO) payload in a mission-specific transfer orbit from Pad 36A. The flight was dedicated to the memory of Mrs. Norma J. Pearce,[29] and it marked the first time in history that an ATLAS/CENTAUR space launch vehicle was used to launch an NRO spacecraft. The payload was processed and encapsulated in the Spacecraft Processing and Integration Facility (SPIF) at Cape Canaveral. Lockheed Martin Astronautics built the launch vehicle, integrated the payload, and flew the mission.[30]

The ATLAS IIA/CENTAUR selected for the mission arrived at Cape Canaveral on 1 December 1997. Technicians erected the ATLAS and its interstage adapter on 4 and 5 January 1998 respectively, and engineers added the CENTAUR upper stage to the launch vehicle on 9 January 1998. Officials completed the Wet Dress Rehearsal (WDR) on 12 January, and engineers mated the spacecraft to the vehicle on 15 January 1998. Following launch certification and RP-1 propellant tanking operations, preparations continued for the first launch attempt on 26 January 1998. During the first launch attempt on 26 January, thick clouds and strong upper level winds conspired to force an unplanned hold at 1816Z, and officials scrubbed the launch at 1845Z. The second launch attempt on 28 January was also scrubbed due to upper level winds at 1642Z. As the launch window opened during the third countdown on 29 January, upper level winds were still too high to launch. Officials decided to extend the built-in hold at T minus 5 minutes. The winds subsided, and the count resumed. Two minutes later (at T minus 3 minutes), the station on Wallops Island experienced a power failure. The crew recycled the count to T minus 5 minutes and waited for the power outage to clear. The count resumed nine minutes later (at 1832Z), and the ATLAS IIA/CENTAUR lifted off Pad 36A at 1837:00.115Z on 29 January 1998. The launch was successful.[31]

ATLAS II/CENTAUR (UHF F-8), AC-132, 16 March 1998

The object of the AC-132 mission was to place the Navy's UHF F-8 communications satellite in an intermediate transfer orbit from Pad 36A. To complete its flight, the spacecraft would use its liquid apogee motor to accelerate to a geosynchronous transfer orbit and final circular orbit. Hughes Aircraft Company, Space and Communications Group (El Segundo, California) built the spacecraft. Astrotech Space Operations in Titusville, Florida, processed the UHF F-8. Lockheed Martin Astronautics integrated the payload to its ATLAS II/CENTAUR vehicle and launched it into space. The spacecraft was one in a series of UHF communications spacecraft being orbited to replace and upgrade the Navy's Fleet Satellite Communications (FLTSATCOM) constellation as well as leased satellites used in the Navy's global communications network. The UHF F-8 had on on-orbit lift expectancy of 14 years.[32]

The ATLAS II/CENTAUR chosen for the mission arrived at Cape Canaveral on 28 January 1998. Technicians erected the ATLAS and its interstage adapter on 3 February, and engineers added the CENTAUR upper stage to the launch vehicle on 4 February 1998. Officials completed the Wet Dress Rehearsal (WDR) on 27 February, and engineers mated the spacecraft to the launch vehicle on 5 March 1998. Following launch certification and RP-1 propellant tanking operations, preparations continued for the scheduled launch on 16 March 1998. During the countdown on the 16th, one of the range's radars did not past slew checks, but officials considered it at least Partially Mission Capable (PMC). Range officials determined the radar met mission requirements, so they committed it at 1913Z to support the launch. The "A" timing string began taking data "hits" about half an hour later, and officials declared it Not Mission Capable (NMC) at 1957Z. Fortunately, String "A" recovered, and authorities declared it Fully Mission Capable (FMC) at 2048Z. There was also a problem with the primary Computer Controlled Launch System (CCLS) in the lower level of Complex 36's blockhouse. It prompted officials to declare a "NO-GO" at 2115Z with an extension of the T minus 5 minutes hold. The problem was resolved, and the count resumed at 2127Z. The ATLAS II/CENTAUR lifted off Pad 36A at 2132:00.340Z on 16 March 1998. The launch was successful.[33]

ATLAS IIA/CENTAUR (UHF F-9), AC-130, 20 October 1998

The object of the AC-130 mission was to place the Navy's UHF F-9 communications satellite in an intermediate transfer orbit from Pad 36A. To complete its flight, the spacecraft would use its liquid apogee motor to accelerate to a geosynchronous transfer orbit and final circular orbit. Hughes Aircraft Company, Space and Communications Group (El Segundo, California) built the UHF F-9, and Astrotech Space Operations in Titusville, Florida, processed the spacecraft. It was integrated to its ATLAS IIA/CENTAUR and launched into space by Lockheed Martin Astronautics. The spacecraft was one in a series of UHF communications spacecraft being orbited to replace and upgrade the Navy's Fleet Satellite Communications (FLTSATCOM) constellation as well as leased satellites used in the Navy's global communications network. The UHF F-9 had on on-orbit lift expectancy of 14 years.[34]

The ATLAS IIA/CENTAUR selected for the launch arrived at Cape Canaveral on 23 July 1998. Technicians erected the ATLAS and its interstage adapter on 3 August, and engineers added the CENTAUR to the launch vehicle on 4 August 1998. After officials completed the Wet Dress Rehearsal on 17 September, engineers erected the spacecraft on 10 October 1998. The vehicle was certified for launch on 12 October 1998. Following RP-1 propellant tanking operations, work continued for the launch on 20 October 1998. There was only one unplanned hold during the countdown on the 20th. It involved a spacecraft reset at 0702Z. The countdown resumed thereafter, and the ATLAS IIA/CENTAUR lifted off Pad 36A at 0719:01.839Z on 20 October 1998. The launch was successful.[35]

ATLAS IIA/CENTAUR (UHF F-10), AC-136, 23 November 1999

The object of the AC-136 mission was to place the Navy's UHF F-10 communications satellite in an intermediate transfer orbit from Pad 36B. Once the transfer orbit was achieved, the spacecraft used its liquid apogee motor to accelerate to a geosynchronous transfer orbit and final circular orbit. Hughes Aircraft Company, Space and Communications Group manufactured the UHF F-10, and Astrotech Space Operations Company processed the spacecraft in Titusville, Florida. Lockheed Martin Astronautics mated the payload to its launch vehicle and launched it into space. The spacecraft was one in a series of UHF communications spacecraft being orbited to replace and upgrade the Navy's Fleet Satellite Communications (FLTSATCOM) constellation as well as leased satellites used in the Navy's global communications network.[36] The UHF F-10 had an on-orbit life expectancy of 14 years.[37]

Technicians erected the ATLAS IIA/CENTAUR on Pad 36B on 23 August 1999 as planned. However, other milestones were delayed about four weeks when a "suspect" CENTAUR engine had to be replaced on the launch vehicle's upper stage on 1 October 1999. Officials completed the Simulated Flight Test and the Wet Dress Rehearsal on 13 and 28 October 1999 respectively. The Navy accepted the new CENTAUR engine on 29 October 1999, and the launch date was moved from 4 to 22 November 1999 to accommodate the delay. Engineers mated the spacecraft to its launch vehicle on 11 November, and officials completed the Composite Electrical Readiness Test on 12 November 1999. Processing continued for the first launch attempt on 21 November 1999.[38]

The mission required two launch attempts. During the countdown on 21 November 1999, there was a 40-minute-long hold at T minus 180 minutes due to high winds. The countdown resumed, but then the ATLAS' main battery failed. The battery was replaced, but it could not be tested in time to meet the launch window.[39] The user scrubbed the launch at 0148Z on 22 November 1999, and officials recycled the countdown for a 23 November launch. The second countdown was far less problematic than the first, but two major problems surfaced before the launch. At 2236Z on the 22nd, the Launch Weather Officer (LWO) declared a "No-Go" for cumulus clouds, and the range remained 'Red' until 0037Z on the 23rd. At 0135Z, Antigua experienced a power outage, and the station's command and telemetry systems were declared Non-Mission Capable (NMC). Fortunately, Antigua was back up around 0203Z, and the countdown proceeded to the launch. The ATLAS IIA/CENTAUR lifted off Pad 36B at 0406:01.086Z on 23 November 1999. The launch was successful.[40]

ATLAS IIA/CENTAUR (DSCS III B8), AC-138, 21 January 2000

As part of the Air Force's effort to upgrade and replace old satellites in the Defense Satellite Communications System (DSCS), the object of the AC-138 mission was to place the DSCS III B8 spacecraft and its Integrated Apogee Boost Subsystem (IABS) in a sub-synchronous transfer orbit from Pad 36A. After the payload separated from the CENTAUR upper stage, the IABS would fire to transfer the payload to near-geosynchronous orbit. Following Spacecraft/IABS separation, the DSCS III would use onboard propulsion systems to transfer itself to circular geosynchronous orbit. Lockheed Martin Missiles and Space built the spacecraft, and the DSCS III was processed in the DSCS Processing Facility (DPF). Lockheed Martin Astronautics built the launch vehicle, integrated the payload and flew the mission.[41]

Technicians erected the ATLAS IIA/CENTAUR on 6 December 1999. Following the Wet Dress Rehearsal (WDR) on 6 January 2000, engineers mated the DSCS III B8 spacecraft to the launch vehicle on 10 January 2000. Officials completed the Composite Electrical Readiness Test (CERT)) on 12 January, and the countdown proceeded as scheduled on 20 January 2000. The ATLAS IIA/CENTAUR lifted off Pad 36A at 0103:01.208Z on 21 January 2000. The launch was successful. In the weeks that followed, the 2,700-pound spacecraft maneuvered to its final position among ten other spacecraft in the DSCS constellation.[42]

ATLAS IIA/CENTAUR (DSCS III B11), AC-140, 20 October 2000

As part of the Air Force's effort to upgrade/replace old satellites in the Defense Satellite Communications System (DSCS) constellation, the object of the AC-140 mission was to place the DSCS III B11 spacecraft and its Integrated Apogee Boost Subsystem (IABS) in a sub-synchronous transfer orbit from Pad 36A. An ATLAS IIA/CENTAUR was selected as the launch vehicle for this mission. According to the flight scenario, the vehicle would fly on an azimuth of 104 degrees, and it would carry the payload to transfer orbit. After the payload separated from the CENTAUR upper stage, the IABS would fire to transfer the payload to near-geosynchronous orbit. Following Spacecraft/IABS separation, the DSCS III would use its own propulsion system to transfer itself to circular geosynchronous orbit approximately 23,230 miles above Earth's surface.[43]

Lockheed Martin Missiles and Space built the DSCS III B11. The spacecraft weighed approximately 2,700 pounds, and it was worth about $200 million. Like the eleven DSCS III satellites already on orbit, the B11 spacecraft was designed to provide uninterrupted secure voice and high rate data communications services to Defense Department users. Unlike most of its predecessors, the B11 was the second of only four DSCS satellites to feature Service Life Enhancement Program (SLEP) improvements for greater reliability and longer service life.[44] The spacecraft was processed in the DSCS Processing Facility (DPF) at Cape Canaveral. Lockheed Martin Astronautics built the launch vehicle, integrated the payload, and flew the mission.[45]

Technicians erected the ATLAS IIA/CENTAUR booster on 28 August 2000. Due to a problem detected in a remote control unit at the factory, Lockheed Martin decided to delay the Simulated Flight Test one day (i.e., until 19 September) to confirm the problem would have no impact on the mission currently underway. In similar fashion, the contractor delayed the Wet Dress Rehearsal for three days (e.g., until 29 September 2000) to remove two Liquid Oxygen (LOX) ducts and install modified ducts after a CENTAUR LOX duct on another vehicle failed vibration testing in the factory earlier in the month. Engineers mated the DSCS III B11 spacecraft to the vehicle on 2 October, but a spacecraft anomaly was detected during the Composite Electrical Readiness Test on 4 October 2000. An investigation revealed the anomaly caused no added risk to the mission. As a result, the spacecraft community decided to fly the vehicle "as is," and Range Safety concurred with the finding on 13 October 2000. The launch was rescheduled from 12 to 19 October 2000, and preparations continued with that date in mind.[46]

The countdown on 18, 19, and 20 October 2000 encountered a number of problems, but none of them led to a scrub. Diego Garcia went Non-Mission Capable (NMC) when it lost communications with Schriever AFB. That outage lasted from 2209Z to 2321Z on 18 October, but the station was ready to support the mission thereafter. In the meantime, the range went 'Red' for weather between 2245Z and 2312Z. At 2342Z, the Flight Control Officer reported a barge heading into the launch hazard area. The incident would have impacted the T-0, but the user asked for a new T-0 to correct a problem with a computer master data file. Ultimately, the T-0 was moved to 0040Z on 20 October. The barge problem was cleared up at 0011Z, and the ATLAS IIA/CENTAUR lifted off Pad 36A at 0040:01.768Z on 20 October 2000. The flight was successful, and the payload was placed in the proper geosynchronous transfer orbit.[47]

ATLAS IIAS/CENTAUR (NRO MLV-11), AC-157, 6 December 2000

The object of the AC-157 mission was to place a National Reconnaissance Office (NRO) classified payload in a "mission specific" transfer orbit from Pad 36A. According to the flight scenario, the ATLAS IIAS/CENTAUR used for the mission would be launched on a flight azimuth of 94 degrees. It would carry the payload to a low Earth parking orbit approximately 9 minutes and 55 seconds after liftoff. Following a coasting period, the CENTAUR upper stage would fire again (e.g., 2nd Burn) about 24 minutes and 19 seconds into the flight. The 2nd burn would provide sufficient energy to carry the CENTAUR and its MLV-11 payload to the mission specific transfer orbit. The spacecraft would separate from the CENTAUR approximately 29 minutes and 37 seconds after liftoff. Lockheed Martin Astronautics was responsible for integrating the payload and launching the ATLAS IIAS/CENTAUR vehicle.[48]

The mission was originally scheduled for launch on 20 March 2000, but a problem with the spacecraft surfaced during processing on 28 February 2000. The MLV-11 payload remained at the Spacecraft Processing & Integration Facility (SPIF) until the problem was corrected. Technicians erected the ATLAS IIAS/CENTAUR on Pad 36A on 27 October 2000. Four Castor IVA solid rocket motors were attached to the vehicle on 30 October (according to schedule), but the Simulated Flight Test had to be re-accomplished on 14 November after some ground support equipment failed during an identical test on 13 November 2000. High winds on 21 November delayed the spacecraft/launch vehicle mating until 22 November 2000, but officials completed the Composite Electrical Readiness Test successfully on 27 November, and the delay had no impact on the launch date (e.g., 5 December 2000). Unfortunately, officials needed more time to accomplish and review "hot fire" testing of the launch vehicle's RL-10 engine after a similar engine at the Pratt & Whitney factory developed a problem. The Launch Readiness Review (LRR) was delayed until 4 December, and the launch was delayed until 6 December 2000.[49]

The countdown on 5/6 December 2000 went well, though there were a couple of significant incidents that deserve comment. At 0105Z on 6 December, the user declared a "No-Go" when a temperature threshold was exceeded. The No-Go was rescinded, but the declaration of the hold prompted Range Safety to send out surveillance helicopters at 0122Z and 0203Z on the 6th. A ship was discovered during the later of the two surveillance flights, and the hold was extended to allow the ship to cross the "box" (i.e. an impact danger area) rather than hold up the ship "any longer than necessary". The ship cleared the box at 0230Z, and the count continued. A radar was down for an array problem for three minutes just before the launch (e.g., 0240-0243Z), but it supported the ATLAS IIAS/CENTAUR flight. First motion was recorded at 0247:01.254Z on 6 December 2000. The MLV-11 payload was placed in the proper geosynchronous transfer orbit about 30 minutes later. The launch was successful.[50]

ATLAS IIAS/CENTAUR (NRO MLV-12), AC-162, 11 October 2001

The object of the AC-162 mission was to place the National Reconnaissance Office's classified MLV-12 payload (NRO MLV-12) into a pre-determined orbit. According to the flight scenario, an ATLAS IIAS/CENTAUR vehicle would lift off Pad 36B on a flight azimuth of 94 degrees. Following the ATLAS IIAS phase of the mission (i.e., the first five minutes of the flight), the CENTAUR upper stage and payload would separate from the core vehicle. The CENTAUR's main engine would fire for approximately four minutes and 39 seconds to inject the MLV-12 into a parking orbit about 10 minutes after lift-off. Following a coasting period, the CENTAUR's second 'burn' would place the MLV-12 in a mission specific transfer orbit. Its mission completed, the CENTAUR upper stage would separate from the spacecraft 29 minutes and 12 seconds after lift-off.[51]

The ATLAS IIAS/CENTAUR chosen for the mission was erected on Pad 36B on 7 August 2001. Processing continued with no impact on the launch date until a crane problem surfaced at the launch site in late August. The Castor IVA solid rocket motors were supposed to be attached to the vehicle on 21 August, but technicians discovered some broken strands on the Torus crane's cabling, and the Castor IVA mate was delayed until the cable could be replaced and load tested. Unfortunately, the crane's upper block bolt heads broke loose during testing in late August, and the crane had to be repaired over the course of the next three weeks. Lockheed Martin's technicians proof-loaded the crane on 21 September, but the crane's secondary brake and upper limit switch both failed during load testing on that date. The parts were replaced and tested about a week later, and the crane supported the Castor IVA mating operation on 29 September 2001. Taken together, the Torus crane problems delayed the launch from 2 October to 11 October 2001, but the rest of Lockheed Martin's preparations went smoothly. Contractor personnel mated the spacecraft to the vehicle on 1 October, and officials completed the Composite Electrical Readiness Test successfully on 2 October 2001. There were no unscheduled holds during the countdown on 10/11 October, and the ATLAS IIAS/CENTAUR lifted off Pad 36B without incident at 0232:00.635Z on 11 October 2001. The vehicle placed the MLV-12 in the proper transfer orbit as planned. This successful launch heralded the Eastern Range's return to operations after about five weeks of Range Standardization and Automation (RSA) upgrades.[52]

ATLAS IIAS/CENTAUR (NROL-1), AC-167, 31 August 2004

The object of the AC-167 mission was to place the National Reconnaissance Office's NROL-1 spacecraft into a proper transfer orbit. The DSCS Processing Facility on the Cape was used to process the "commercially procured" Defense Department payload. Lockheed Martin Space Systems Company was responsible for producing the launch vehicle, integrating the mission, and launching the ATLAS IIAS/CENTAUR. The mission marked the final flight of the ATLAS IIAS/CENTAUR launch program, and it was the final ATLAS/CENTAUR mission to be launched from Pad 36A.[53]

According to the flight scenario, the ATLAS IIAS/CENTAUR would rise vertically and roll into a flight azimuth of 45 degrees with an inclination of 58 degrees. Following staging sequences, the spacecraft would be released from the vehicle approximately 73 minutes and 49 seconds after lift-off. Officials scheduled the launch for 24 June 2004, but the date slipped to 1 July before becoming "indefinite" pending completion of a range flight profile analysis and an investigation of a CENTAUR remote control unit failure at the factory in June 2004. By 6 July, officials had rescheduled the launch for 28 July 2004. In the meantime, the simulated flight was completed on 8 June, and officials completed the Wet Dress Rehearsal (WDR) on 15 June 2004. Engineers mated the spacecraft to the launch vehicle on 16 June, and officials conducted the Combined Electrical Readiness Test on 23 June 2004. Vehicle problems continued to affect the launch date, however, and the mission went "indefinite" until officials resolved existing discrepancies well enough to choose 31 August 2004 as the new launch date. By the end of July a new launch date of 27 August had been proposed, and officials approved it as the new launch date on 2 August 2004.[54]

Unfortunately, weather problems began to intrude in early August. A lightning storm knocked out power to Complex 36 on 3 August 2004, but backup systems performed well with "virtually no interruption" in service. The Mobile Service Tower on Pad 36A suffered a direct lightning strike on 5 August, so officials retested critical ground power systems and safety circuits on 6 August 2004 to ensure their proper operation. Hurricane preparations in mid-August were required due to the threat of Hurricane CHARLEY, but all facilities "faired exceptionally well" after CHARLEY came ashore from the Gulf of Mexico and proceeded through Central Florida toward Daytona Beach before exiting into the Atlantic Ocean on 15 August 2004.[55]

Technicians fueled the ATLAS IIAS with RP-1 (highly refined kerosene) on 19 August, but a set of special radio frequency tests on 21 and 22 August delayed nitrogen tetroxide loading operations until the 23rd. Technicians had to replace a CENTAUR battery on 26 August due to a low voltage reading, and the user requested the countdown be delayed until 28 August to resolve any lingering concerns over the battery problem. As events turned out, officials had to scrub the launch on the 28th due to lack of liquid oxygen propellant. Weather intruded during the countdown on the 29th, and officials scrubbed the launch at 2325Z on the 29th due to an "anvil cloud" rule violation that rendered the range 'Red.' They also scrubbed another countdown at 2343Z on the 30th due to weather violations. Finally, with no unplanned holds during the countdown on the 31st, the ATLAS IIAS/CENTAUR and its NROL-1 payload lifted off Pad 36A at 2317:00.717Z on 31 August 2004. The launch was successful.[56]

ATLAS I/CENTAUR and ATLAS II/CENTAUR Commercial Space Operations

ATLAS IIAS/CENTAUR (TELSTAR 4), AC-108, 16 December 1993

The object of the AC-108 mission was to boost the TELSTAR 4 communications satellite[57] into an elliptical transfer orbit with an apogee of 19,409 nautical miles. General Dynamics used its first commercial ATLAS IIAS/CENTAUR vehicle for the mission, and the company prepared the booster for launch from Pad 36B. The CENTAUR upper stage arrived at the Cape on 14 September 1993, and technicians and engineers erected the ATLAS IIAS/CENTAUR stages by the end of the month. Officials completed the launch vehicle simulated flight test in early November 1993, and they conducted the Wet Dress Rehearsal a few days later. Following encapsulation of the TELSTAR 4 spacecraft, engineers mated the payload to the launch vehicle in early December. The vehicle was prepared for launch on 14 December 1993.[58]

Officials scrubbed the first launch attempt on 14 December around 2112Z when high winds prevented the safe retraction of the Mobile Service Tower. The countdown was recycled for another attempt on 15 December, and it proved to be anything but uneventful: Satellite communications with Antigua were partially out at 1613Z on the 15th, but trouble-shooting relieved the problem several hours before the launch. A Distant Object Attitude Measuring Systems (DOAMS) had timing problems, and a portable timing unit had to be brought onsite to allow the DOAMS to support the launch. At 2030Z Bermuda's telemetry data was unstable due to a software problem, and officials pulled Bermuda's telemetry system out of the range solution. With Bermuda's telemetry out of the picture, Antigua's telemetry became mandatory for the mission. As luck would have it, the telemetry systems controller began getting incorrect data from Antigua around 2100Z. Fortunately, the source of the problem was a crossed connection — it was quickly set right, and Antigua's telemetry was declared Fully Mission Capable (FMC) at 2131Z. In the meantime, the optics mother computer went out at 2115Z due to a bad timing card. Technicians reseated the card, and the system came back online by 2124Z. The MARSS weather terminal locked up two minutes later, but the problem was quickly traced to KSC's VAX computer memory bank. Once the VAX was rebooted, the problem cleared. The MARSS was ready to support the mission at 2155Z. The Cape's Command Message Encoder Verifier (CMEV) was partially out at 2305Z for low power, but trouble-shooting brought it back by 2327Z. Antigua's command system began "taking hits" three minutes later, but engineers traced the problem to the Cape's CMEV, and officials cleared the command system by 0016Z on the 16th. A camera site failed to support the mission due to an encoder problem at 2345Z on the 15th. A different camera site was substituted to cover the commitment. Despite all the problems mentioned, only one — a faulty indication on the Test Conductor's console in the blockhouse — caused any extension of the planned holds for the launch. The Test Conductor was not getting a "hold down" indication on his console, and the final built-in hold was extended 18 minutes to deal with that problem. The countdown resumed at 0035Z, and the Atlas IIAS/CENTAUR lifted off Pad 36B at 0040:01.884Z on 16 December 1993. The launch was successful.[59]

ATLAS IIA/CENTAUR (DIRECTV), AC-107, 3 August 1994

The object of the AC-107 mission was to place the DIRECTV commercial communications spacecraft in a near geosynchronous transfer orbit. The spacecraft was the second of two Direct Broadcasting Satellites (DBS) launched into space to deliver more than 150 channels of TV entertainment to homes throughout North America.[60] Technicians erected the ATLAS IIA booster/sustainer and interstage adapter on Pad 36A on 31 May 1994. Engineers erected the CENTAUR upper stage on 1 June 1994. Following the Wet Dress Rehearsal and spacecraft/vehicle mating operation in July, technicians and engineers prepared the vehicle for launch on 28 July 1994. Unfortunately, the launch was delayed on that date due to a bad Inertial

Navigation Unit power supply. Officials rescheduled the launch for 3 August 1994. Following an uneventful countdown, the vehicle lifted off Pad 36A at 2357:02.202Z on that date. The spacecraft achieved super-synchronous orbit (as planned), and the mission was a success. AC-107 was the first commercial ATLAS mission launched from Pad 36A. (Previous commercial ATLAS missions had been launched from Pad 36B.)[61]

ATLAS IIAS/CENTAUR (INTELSAT VII), AC-111, 6 October 1994

The object of the AC-111 mission was to place an INTELSAT VII communications satellite (INTELSAT 703) into a sub-transfer orbit so the spacecraft's bipropellant propulsion system could lift the payload to transfer orbit and final geosynchronous orbit. The most powerful ATLAS vehicle of the time — the ATLAS IIAS/CENTAUR — was chosen to lift the 8,118-pound INTELSAT VII payload into space.[62] Cape Canaveral received the launch vehicle on 20 July 1994. Technicians erected the ATLAS IIAS booster/sustainer on Pad 36B on 23 July, and they raised the interstage on 25 July 1994. Engineers erected the CENTAUR upper stage on 29 July 1994. Following the Wet Dress Rehearsal on 30 and 31 August 1994, technicians installed four Castor IVAs on the vehicle on 7 and 10 September. Engineers mated the spacecraft to the launch vehicle in mid-September, and the ATLAS IIAS/CENTAUR was prepared for launch on 6 October 1994. The countdown on 6 October was delayed about 42 minutes (from 0548Z to 0630Z) to ensure Bermuda's telemetry was available in case Antigua's telemetry went down. The countdown proceeded smoothly thereafter, and the launch vehicle lifted off Pad 36B at 0635:02.358Z on 6 October 1994. The spacecraft separated from the vehicle 28 minutes later, and it entered a very high (24,046 nautical mile) orbit to give the $209 million satellite additional months of service on-orbit. The launch was successful.[63]

ATLAS IIA/CENTAUR (ORION), AC-110, 29 November 1994

The object of the AC-110 mission was to place the ORION communications spacecraft in a super-synchronous transfer orbit. Following spacecraft separation, ORION's first burn was expected to place the spacecraft in a highly elliptical orbit. The ORION spacecraft weighed approximately 5,200 pounds, and it was nearly cubical, measuring approximately 2 meters along each edge. In its deployed configuration, the ORION's solar arrays spanned 22.0 meters. British Aerospace built the spacecraft for ORION Atlantic, a partnership consisting of ORION Network Systems and several British, Canadian, Italian, and Japanese firms. The satellite provided communications services for North America and Europe. ORION's transponders were divided equally between the two continents. Six of the spacecraft's 34 solid-state transponders operated at 36 MHz, and the other transponders operated at 54 MHz. The satellite had an operational life expectancy of 12 years.[64]

The ATLAS IIA vehicle chosen for the mission arrived at Cape Canaveral on 25 September 1994. Technicians erected the ATLAS booster/sustainer on Pad 36A on 28 September, and they mounted the interstage adapter on the vehicle on 29 September 1994. Following the CENTAUR engine nozzle installation on 3 October, engineers erected the CENTAUR upper stage on 10 October 1994. The Wet Dress Rehearsal and spacecraft mating followed shortly thereafter, and the vehicle was readied for a launch attempt on 22 November 1994. There was one unplanned hold during the countdown on 22 November. A "sticking" liquid oxygen fill and drain valve caused the delay, and the unplanned hold lasted 41 minutes. The countdown resumed at 1100Z, but officials stopped it five minutes later when the ATLAS IIA's engines failed to start. The failure was attributed to launch control computers. Officials scrubbed the launch at 1104:54Z on the 22nd, and they rescheduled the launch for 29 November 1994. The countdown on the 29th was uneventful, and the vehicle lifted of Pad 36A at 1021:01.927Z on 29 November 1994. The launch was successful. Based on a telemetry report from Advanced Range Instrumentation Aircraft (ARIA) supporting the mission, the spacecraft achieved a successful elliptical orbit. The ORION's engines fired later on, injecting the spacecraft in a super-synchronous orbit.[65]

ATLAS IIAS/CENTAUR (INTELSAT VII), AC-113, 10 January 1995

The object of the AC-113 mission was to place an INTELSAT VII commercial communications satellite (INTELSAT 704-2) into a sub-transfer orbit so the spacecraft's bi-propellant propulsion system could lift the payload into transfer orbit and final geosynchronous orbit. Lockheed Martin's most powerful ATLAS vehicle at the time — the ATLAS IIAS — was used to boost the 8,118-pound INTELSAT VII payload into space. Space Systems/Loral built the spacecraft under a contract with the International Telecommunications Satellite (INTELSAT) Organization. The satellite provided voice, video, and data transmission services for INTELSAT member nations. The INTELSAT VII had an operational life expectancy of 10.9 years. The Astrotech Space Operations Company processed spacecraft through the company's facilities in Titusville, Florida. Martin Marietta Space Systems (a division of Lockheed Martin) was responsible for production of the launch vehicle, mission integration, total system performance, launch, and flight operations.[66]

The ATLAS IIAS/CENTAUR vehicle arrived at Cape Canaveral on 29 October 1994. Technicians erected the ATLAS on Pad 36B on 1 November, and technicians and engineers added the interstage adapter and CENTAUR upper stage on 2 and 3 November 1994 respectively. Officials conducted the launch vehicle simulated flight test on 3 December, and they completed the Wet Dress Rehearsal on 6 December 1994. The vehicle's solid rockets were installed between 6 and 14 December 1994, and engineers mated the INTELSAT VII to the launch vehicle on 22 December 1994. Following the certification for launch on 3 January 1995, preparations continued for the launch on January 10th. There were no unplanned holds during the countdown, and the vehicle lifted off Pad 36B at 0618:02.485Z on 10 January 1995. The launch was successful, and the INTELSAT joined a constellation of more than 20 communications satellites providing services to more than 200 countries and territories around the world.[67]

ATLAS IIAS/CENTAUR (INTELSAT VII), AC-115, 22 March 1995

The third in a series of three INTELSAT VII communications satellites was prepared for launch from Cape Canaveral in March 1995. Like its siblings, the INTELSAT 704-3 weighed 8,118 pounds, and it had an on-orbit life expectancy of 10.9 years. It would join the constellation of 24 INTELSAT spacecraft providing communications services to more than 200 INTELSAT member nations around the world.[68]

The mission's ATLAS IIAS/CENTAUR vehicle arrived at Cape Canaveral on 15 January 1995. Technicians erected the ATLAS and its interstage adapter on 19 January, and engineers raised the CENTAUR upper stage on 21 January 1995. Following the launch vehicle simulated flight test on 15 February, officials conducted the Wet Dress Rehearsal on 22 February 1995. All four solid rocket motors were installed on the vehicle by the end of February, and engineers mated the INTELSAT VII to the launch vehicle on 12 March 1995. Officials completed launch certification on 14 March, and preparations continued for a launch on 21 March 1995. Unfortunately, the launch attempt on 21 March was scrubbed after a problem emerged during the T minus 5 minutes built-in hold. The POGO suppressor and the airborne fill/drain valve were not working properly. Despite troubleshooting, the problem could not be resolved in time, and officials scrubbed the mission at 0652Z. The count was recycled 24 hours, and the countdown got underway again at 2218Z on 21 March. This time there were no unscheduled holds, and the ATLAS IIAS/CENTAUR lifted off Pad 36B at 0618:01.427Z on 22 March 1995. The launch was successful.[69]

ATLAS IIA/CENTAUR (MSAT), AC-114, 7 April 1995

The object of the AC-114 mission was to launch the MSAT communications spacecraft into the proper geosynchronous transfer orbit. A commercial ATLAS IIA/CENTAUR vehicle was chosen to support the flight. Following the payload's separation from the vehicle, the spacecraft's onboard propulsion system would boost the MSAT into its final geostationary orbit. Hughes Space and Communications, El Segundo, California, built the MSAT. The Astrotech Space Operations Company processed and encapsulated the payload in Titusville, Florida. The MSAT was designed to provide mobile telephone service to North America by means of two elliptical graphite mesh antennas stowed around the body of the spacecraft. Once the MSAT was on-orbit, a cutting device would be used to cut a restraining band and allow the antennas to unfurl. Measuring 6.7 x 4.9 meters, the antennas were advertised as the largest ever deployed on a commercial communications satellite.[70]

The ATLAS IIA/CENTAUR vehicle arrived at Cape Canaveral on 12 February 1995. Technicians erected the ATLAS IIA and interstage adapter on Pad 36A on 14 and 15 February respectively. Engineers installed the CENTAUR upper stage on 17 February. Officials completed the launch vehicle simulated flight test on 3 March, and spacecraft encapsulation was completed at Astrotech on 20 March 1995. Following the Wet Dress Rehearsal on 23 March, engineers mated the spacecraft to the vehicle on 25 March 1995. Officials completed launch certification on 28 March, and preparations continued for a launch on 5 April 1995. During the countdown on 5 April high-level winds made the launch unlikely, and officials scrubbed the launch at 2047Z on that date. The countdown was recycled 48 hours for the second launch attempt. The countdown on 7 April went smoothly with no unscheduled holds. The ATLAS IIA/CENTAUR lifted off Pad 36A at 2347:00.231Z on 7 April 1995. The launch was successful.[71]

ATLAS IIAS/CENTAUR (JCSAT), AC-117, 29 August 1995

The object of the AC-117 mission was to place the JCSAT, a Japanese communications satellite, into a geosynchronous transfer orbit. Hughes Space & Communications built the JCSAT under a contract with Japan Satellite Systems Inc. The 6,786-pound spacecraft was designed to provide communications services for Asia, Australia, and islands in the Pacific Ocean for 12 years. The Astrotech Space Operations Company in Titusville, Florida, processed the JCSAT, and Lockheed Martin handled production of the ATLAS IIAS/CENTAUR vehicle, mission integration, launch, and flight operations.[72]

The ATLAS IIAS/CENTAUR selected for the mission arrived at Cape Canaveral on 27 June 1995. Technicians erected the ATLAS II on Pad 36B on 29 June, and they added the interstage adapter on 30 June 1995. Engineers raised the CENTAUR upper stage on 5 July 1995. Following the launch vehicle simulated flight test on 29 July, officials completed the Wet Dress Rehearsal on 7 August 1995. Technicians installed the last of the vehicle's four solid rocket motors on 12 August, and engineers mated the JCSAT to the launch vehicle on 16 August 1995. Officials conducted the launch certification on 18 August, and preparation continued for the countdown on 28 August 1995. There were two unplanned five-minute holds during the countdown on 28 August. Weather and a delay in hydrogen or oxygen tanking operations contributed to both of those holds. Ultimately the countdown resumed, and the ATLAS IIAS/CENTAUR lifted off Pad 36B at 0053:01.722Z on 29 August 1995. The launch was successful.[73]

ATLAS IIA/CENTAUR (GALAXY IIIR), AC-120, 15 December 1995

The object of the AC-120 mission was to place the GALAXY IIIR communications satellite into a geosynchronous transfer orbit. The spacecraft's own propulsion system would deliver the payload to final geostationary orbit. Hughes Space and Communications, El Segundo, California, built the spacecraft. The GALAXY IIIR was designed to provide C-Band programming to the Continental U.S. and the Caribbean. The satellite also carried a Ku-Band antenna to deliver DIRECTV entertainment programs to Mexico, the Caribbean, South America, and Central America.[74]

The ATLAS IIA/CENTAUR vehicle was delivered to Cape Canaveral on 18 October 1995. Technicians erected the ATLAS and its interstage adapter on Pad 36A on 1 November, and engineers raised the CENTAUR upper stage on 2 November 1995. Following the simulated launch vehicle flight test on 27 November, officials completed the Wet Dress Rehearsal on 29 November 1995. Workers encapsulated the spacecraft encapsulation on 1 December, and engineers mated the payload to the launch vehicle on 4 December 1995. The launch certification was completed on 6 December, and preparations continued for the launch on 15 December 1995. There were no unscheduled holds during the countdown on 14 December, and the ATLAS IIA/CENTAUR lifted off Pad 36A at 0023:01.151Z on 15 December 1995. The launch was successful.[75]

ATLAS IIAS/CENTAUR (PALAPA-C1), AC-126, 1 February 1996

The object of the AC-126 mission was to place the PALAPA-C1 communications satellite into a super-synchronous transfer orbit. This, in turn, would permit the spacecraft's onboard propulsion system to lift the payload into final geostationary orbit at 113 degrees east longitude. The PALAPA-C1 would provide C-Band and Ku-Band communications over an area extending from Iran and Vladivostok to Australia and New Zealand. Lockheed Martin's most powerful ATLAS vehicle of the time — the ATLAS IIAS/CENTAUR — was used to boost the PALAPA-C1 into space. Hughes Space and Communications built the satellite in El Segundo, California under a contract with the Satelindo organization. The Astrotech Space Operations Company processed the spacecraft through the company's facilities in Titusville, Florida. Martin Marietta Space Systems (a division of Lockheed Martin) was responsible for production of the launch vehicle, mission integration, total system performance, launch, and flight operations. The AC-126 mission marked the debut of a Tracking and Data Relay Satellite System (TDRSS) telemetry link. For flight validation purposes, the new link was loaded "piggyback" with a traditional 2202.5 MHz pulse-code modulation/frequency modulation (PCM/FM) telemetry link. The Eastern Range was configured to support both downlinks.[76]

The ATLAS IIAS/CENTAUR vehicle arrived at Cape Canaveral on 30 November 1995. Technicians erected the ATLAS and its interstage adapter on Pad 36B on 9 December, and engineers added the CENTAUR upper stage to the launch vehicle on 10 December 1995. Officials completed the Wet Dress Rehearsal on 13 January 1996. Technicians installed the vehicle's four solid rocket boosters between 16 and 19 January 1996. Engineers mated the PALAPA-C1 spacecraft to the launch vehicle on 22 January 1996. Following launch certification and RP-1 propellant tanking operations, preparations continued for the launch on 1 February 1996. There was one unplanned 25-minute hold during the countdown. One of two Operations Safety Managers (OSMs) failed to receive proper indications of an enabled ignition switch on his status panel at 0049Z, and the count was recycled to T minus five minutes. The countdown resumed, and the vehicle lifted off Pad 36B smoothly at 0115:01.263Z on 1 February 1996. The launch was successful.[77]

ATLAS IIA/CENTAUR (INMARSAT-3), AC-122, 3 April 1996

The object of the AC-122 mission was to place the first INMARSAT-3 communications satellite into a proper geosynchronous transfer orbit. This would allow the spacecraft to boost itself into final geostationary orbit via its onboard propulsion system. Lockheed Martin Astro Space manufactured the INMARSAT-3 in East Windsor, New Jersey. Astrotech Space Operations in Titusville, Florida processed the spacecraft. Lockheed Martin Astronautics was responsible for integrating the payload and launching the ATLAS IIA/CENTAUR vehicle.[78]

The ATLAS IIA/CENTAUR vehicle arrived at Cape Canaveral on 9 February 1996. Technicians erected the ATLAS and its interstage adapter on Pad 36A on 16 February, and engineers added the CENTAUR upper stage to the launch vehicle on 19 February 1996. Officials completed the Wet Dress Rehearsal on 14 March, and engineers mated the INMARSAT-3 spacecraft to the launch vehicle on 21 March 1996. Following launch certification and RP-1 propellant tanking operations, preparations continued for a launch attempt on 1 April 1996. Unfortunately, low-level winds on 1 April prompted the Launch Director to delay the rollback of the Mobile Service Tower. This resulted in an unscheduled hold at 2004Z, and officials eventually scrubbed the launch for low level winds at 2206Z. The countdown was recycled 24 hours to 2 April, but, once again, low level winds prompted an unscheduled hold at 2002Z. The launch was scrubbed at 2312Z on 2 April 1996. Finally, the countdown on 3 April proved successful. Through one of the range's surveillance helicopters was diverted at 2135Z to support a Coast Guard rescue mission, Another helicopter remained available, and the countdown proceeded smoothly to the ATLAS IIA/CENTAUR's lift-off from Pad 36A at 2301:00.792Z on 3 April 1996. The launch was successful.[79]

ATLAS I/CENTAUR (SAX), AC-78, 30 April 1996

The object of the AC-78 commercial launch operation was to place the X-Ray Astronomy Satellite (SAX) into a 600-kilometer-high circular orbit with an inclination between zero and five degrees to the equator. Alenia Spazio SPA of Turin, Italy built the spacecraft. The mission was a sponsored by the Agenzia Spaziale Italiana (ASI) and the Netherlands Agency for Space Programs (NIVR). The SAX spacecraft contained five scientific instruments and their support systems to observe celestial X-rays sources in the 0.1 to 300 KeV energy band. Once the SAX was on-orbit, it would relay almost 450 megabytes of data with each revolution around the Earth to a ground station in Singapore. The ASI maintained an operational control center near Rome, and that facility was linked to Singapore via communications satellite. Prior to its arrival at Pad 36B, the SAX spacecraft was processed by Astrotech Space Operations in Titusville, Florida. Lockheed Martin Astronautics was responsible for integrating the payload at the launch pad and launching the vehicle.[80]

The ATLAS I/CENTAUR arrived at Cape Canaveral on 27 February 1996. Workers erected the ATLAS and its interstage adapter on Pad 36B on 5 and 6 March respectively, and engineers added the CENTAUR upper stage to the launch vehicle on 7 March 1996. Officials completed the Wet Dress Rehearsal on 9 April, and engineers mated the SAX to the launch vehicle on 17 April 1996. Following launch certification on 19 April and propellant tanking operations a few days later, final preparations continued for the launch on 30 April 1996. On 30 April there was one unscheduled hold at 0131Z for low level winds. Officials continued the range count at 0201Z, and the space launch vehicle lifted off Pad 36B at 0431:01.336Z on 30 April 1996. The launch was successful, and it reportedly marked the 100th launch of an ATLAS/CENTAUR vehicle.[81]

ATLAS IIA/CENTAUR (GE-1), AC-123, 8 September 1996

The object of the AC-123 mission was to place the GE-1 commercial communications satellite in a super-synchronous transfer orbit. Using onboard propellants, the spacecraft's Liquid Apogee Engine (LAE) would boost the payload into its final geostationary orbit. Lockheed Martin Astrospace built the GE-1 spacecraft in East Windsor, New Jersey. Astrotech Space Operations processed the spacecraft in Titusville, Florida, and Lockheed Martin Astronautics mated the payload the launched the vehicle from Pad 36B. The flight was scheduled to lift off on 30 August 1996. Unfortunately, the launch was delayed when the manufacturer of the spacecraft's onboard computer announced that his people had found a problem in an identical onboard computer scheduled for a later mission. The problem delayed the launch until 8 September to allow the unit to be retested and reinstalled.[82]

The ATLAS IIA/CENTAUR arrived at Cape Canaveral on 13 June 1996. Technicians erected the ATLAS and its interstage adapter on 8 July, and engineers added the CENTAUR upper stage to the launch vehicle on 9 July 1996. Officials completed the Wet Dress Rehearsal on 30 July, and engineers mated the spacecraft to the launch vehicle on 8 August 1996. Following launch certification and RP-1 propellant tanking operations and the scheduling delay mentioned earlier, preparations continued for the launch on 8 September 1996. There were no unscheduled holds during the countdown, and the vehicle lifted off Pad 36B at 2149:01.000Z on 8 September 1996. The launch was successful.[83]

ATLAS IIA/CENTAUR (HOT BIRD 2), AC-124, 21 November 1996

The object of the AC-124 mission was to place the HOT BIRD 2 communications satellite in a transfer orbit that would enable the spacecraft to inject itself into final geostationary orbit. Matra Marconi Space built the HOT BIRD 2, and the spacecraft was processed and encapsulated by Astrotech Space Operations in Titusville, Florida. Lockheed Martin Astronautics was responsible for mating the payload to the ATLAS IIA/CENTAUR as well as launching the mission from Pad 36B. The European Telecommunications Satellite (EUTELSAT) Organization owned and operated the HOT BIRD 2, and the satellite would be joining two other orbiting satellites as part of a five-satellite network to be completed in 1998. The satellite was launched to provide cable television and radio services to customers throughout Europe, the Mediterranean, and the Middle East. The HOT BIRD 2 had an operational life expectancy of 15 years.[84]

The ATLAS IIA/CENTAUR arrived at Cape Canaveral on 11 September 1996. Technicians erected the ATLAS and its interstage adapter on Pad 36B on 17 September, and engineers added the CENTAUR upper stage to the launch vehicle on 18 September 1996. Officials completed the Wet Dress Rehearsal on 29 October, and engineers mated the spacecraft to the launch vehicle on 4 November 1996. Following launch certification and RP-1 propellant tanking operations, preparations continued for the first launch attempt on 13 November 1996. Unfortunately, the first launch attempt was frustrated by ground winds in excess of 23 knots. At 1738Z on the 13th, the user decided to move the T minus 105 minutes Built In Hold (BIH) to T minus 135 minutes and extend the hold to delay Mobile Service Tower removal until the wind died down. Unfortunately, the wind did not cooperate. After a delay of one hour and 25 minutes, the user ran out of launch window time, and officials scrubbed the launch at 1938Z. The launch was recycled 24 hours. Grounds winds became a problem again during the countdown on 14 November. At 1813Z, the user decided to extend the BIH at T minus 135 minutes and wait for the wind to die down. There was also an indication of a possible leak in the RP-1 (propellant) ground start system. Eventually, the user ran out of time, and officials declared the scrub at 1938Z for the RP-1 leak indication. The third and final countdown got underway on 21 November, and there were no unscheduled holds this time. The ATLAS IIA/CENTAUR lifted off Pad 36B at 2047:01.187Z on 21 November 1996. The launch was successful.[85]

ATLAS IIA/CENTAUR (INMARSAT-3 F3), AC-129, 18 December 1996

The object of the AC-129 mission was to place the second INMARSAT-3 communications satellite[86] in a transfer orbit that would enable the spacecraft to boost itself into final geostationary orbit. Lockheed Martin Astro Space built the INMARSAT-3 in East Windsor, New Jersey. Astrotech Space Operations in Titusville, Florida, processed the spacecraft. Lockheed Martin Astronautics was responsible for integrating the payload and launching the ATLAS IIA/CENTAUR vehicle.[87]

The ATLAS IIA/CENTAUR vehicle arrived at Cape Canaveral on 30 October 1996. Technicians erected the ATLAS and its interstage adapter on Pad 36A on 7 and 8 November respectively, and engineers added the CENTAUR upper stage to the launch vehicle on 9 November 1996. Officials completed the Wet Dress Rehearsal on 5 December, and engineers mated the spacecraft to the launch vehicle on 10 December 1996. Following launch certification and RP-1 propellant tanking operations, preparations continued for the launch on 18 December. There were no unscheduled holds during the countdown, and the ATLAS IIA/CENTAUR lifted off Pad 36A at 0157:00.872Z on 18 December 1996. The launch was successful.[88]

ATLAS IIAS/CENTAUR (JCSAT-4), AC-127, 17 February 1997

The object of the AC-127 mission was to place the JCSAT-4 communications satellite into transfer orbit from Pad 36B. Hughes Space & Communications built the JCSAT under contract to Japan Satellite Systems, Inc. The spacecraft weighed 6,810 pounds, and it was designed to provide commercial communications services for Asia, Australia, and islands in the Pacific Ocean. The Astrotech Space Operations Company was contracted to process the spacecraft through the company's facilities in Titusville, Florida. Lockheed Martin was responsible for mission integration, production of the launch vehicle, and launch operations.[89]

The ATLAS IIAS/CENTAUR arrived at Cape Canaveral on 2 December 1996. Technicians erected the ATLAS and its interstage adapter on 9 and 10 December respectively, and engineers raised the CENTAUR upper stage on 11 December 1996. Officials completed the Wet Dress Rehearsal on 23 January 1997, and engineers mated the spacecraft to the launch vehicle on 4 February 1997. Following launch certification and RP-1 propellant tanking operations, preparations continued for the launch on 17 February 1997. There were two unplanned holds during the countdown on 16/17 February. Officials called the first hold for thick clouds and heavy ground winds at 2207Z on the 16th. The hold was extended until 2300Z. Similar cloud conditions prompted the second hold at 0125Z on February 17th. Fortunately, the situation improved, and the count resumed at 0137Z. The ATLAS IIAS/CENTAUR lifted off Pad 36B at 0142:01.580Z on 17 February 1997. The launch was successful.[90]

ATLAS IIA/CENTAUR (TEMPO II), AC-128, 8 March 1997

The object of the AC-128 mission was to place the TEMPO II spacecraft in a sub-synchronous transfer orbit from Pad 36A. Space Systems/Loral built the TEMPO II for TEMPO Satellite, Inc., a subsidiary of Tele-Communications, Inc. Astrotech Space Operations (ASO) was contracted to process the satellite at ASO's facilities in Titusville, Florida. Lockheed Martin Astronautics was responsible for mission integration, production of the launch vehicle, and launch operations. TCI Satellite Entertainment purchased the TEMPO II to provide direct-to-home television service throughout the United States, including Alaska, Hawaii, and Puerto Rico.[91]

The ATLAS IIA/CENTAUR vehicle arrived at Cape Canaveral on 8 January 1997. Technicians erected the ATLAS and its interstage adapter on 14 and 15 January respectively, and engineers raised the CENTAUR upper stage on 16 January 1997. After the Wet Dress Rehearsal on 11 February, engineers mated the spacecraft to the launch vehicle on 17 February 1997. Officials certified the vehicle for launch on 19 February 1997. Following RP-1 propellant tanking operations, work continued for the first launch attempt on 5 March 1997. Unfortunately, four attempts were required on four separate days before the ATLAS IIA/CENTAUR vehicle lifted off Pad 36A. Officials scrubbed the first launch attempt at 2350Z on 4 March due to a defective ATLAS battery. The launch contractor scrubbed the second countdown at 0310Z on 6 March due to problems at the spacecraft command center in Palo Alto, California. The launch attempt on March 7 was scrubbed at 0450Z due to high ground winds. Fortunately, there were no unplanned holds during the countdown on 8 March, and the ATLAS IIA/CENTAUR lifted off Pad 36A at 0601:01.000Z on 8 March 1997. The launch was successful.[92]

ATLAS IIAS/CENTAUR (SUPERBIRD-C), AC-133, 28 July 1997

The object of the AC-133 mission was to place the SUPERBIRD-C communications satellite in a transfer orbit from Pad 36B. Hughes Space & Communications built the spacecraft under a contract to Space Communications Corporation (SCC). The SUPERBIRD-C spacecraft weighed 6,810 pounds, and it had an operational life expectancy of 13 years. The satellite's 24 Ku-Band channels were designed to provide commercial communications services for Japan, much of Asia, Australia, and islands in the Pacific Ocean. The payload was processed through Astrotech Space Operations' facilities in Titusville, Florida. Lockheed Martin Astronautics produced the launch vehicle, integrated the mission, and flew it.[93]

The ATLAS IIAS/CENTAUR vehicle arrived at Cape Canaveral on 7 May 1997. Technicians erected the ATLAS and its interstage adapter on 13 and 14 May respectively, and engineers raised the CENTAUR upper stage on 15 May 1997. After the Wet Dress Rehearsal on 17 June, engineers mated the spacecraft to the launch vehicle on 27 June 1997. Officials certified the vehicle for launch on 30 June 1997. Following RP-1 propellant tanking operations, preparations continued for the first launch attempt on 26 July 1997. Officials scrubbed the first launch attempt at 0111:01Z on the 26th due to a booster hold down pressure regulator problem. The second launch (set for 27 July) was scrubbed at 2308:00Z on 26 July due to ground winds and lightning. The final attempt on 28 July 1997 experienced no unplanned holds, and the ATLAS IIAS/CENTAUR lifted off Pad 36B at 0115:01.109Z on that date. The launch was successful.[94]

ATLAS IIAS/CENTAUR (GE-3), AC-146, 4 September 1997

The object of the AC-146 mission was to place the GE-3 commercial communications satellite in a geosynchronous transfer orbit from Pad 36A. Lockheed Martin Missiles and Space built the spacecraft under contract to Lockheed Martin Telecommunications. Lockheed Martin Astronautics was responsible for mission integration and launch/flight operations, and Astrotech

Space Operations processed the spacecraft at facilities in Titusville, Florida before the payload was transported to the launch pad.[95]

The ATLAS IIAS/CENTAUR vehicle arrived at Cape Canaveral on 5 June 1997. Technicians erected the ATLAS and its interstage adapter on 9 and 16 June respectively, and engineers raised the CENTAUR upper stage on 17 June 1997. After the Wet Dress Rehearsal on 7 August, engineers mated the spacecraft to the launch vehicle on 14 August 1997. Officials certified the vehicle for launch on 18 August 1997. Following RP-1 propellant tanking operations, preparations continued for the first launch attempt on 3 September 1997. There were two unscheduled holds during the countdown on the 3rd. The first hold was called at 1202Z after incorrect pressure was detected in the CENTAUR's liquid oxygen tank. Officials picked up the count at 1223Z, but they were forced to call another unscheduled hold at 1227Z for incorrect pressure in the CENTAUR's liquid oxygen tank. The launch was scrubbed at 1228Z, and the count was recycled 24 hours. Fortunately, there were no unplanned holds during the second launch attempt, and the ATLAS IIAS/CENTAUR lifted off Pad 36A at 1203:00.421Z on 4 September 1997. The launch was successful.[96]

ATLAS IIAS/CENTAUR (ECHOSTAR III), AC-135, 5 October 1997

The object of the AC-135 mission was to place the ECHOSTAR III commercial communications satellite in a geosynchronous transfer orbit from Pad 36B. Lockheed Martin Missiles and Space built the 8,020-pound spacecraft under contract to EchoStar. The satellite was processed at Astrotech's facilities in Titusville, Florida. Lockheed Martin Astronautics built the vehicle, integrated the payload, and flew the mission.[97]

The ATLAS IIAS/CENTAUR vehicle arrived at Cape Canaveral on 29 July 1997. Technicians erected the ATLAS and its interstage adapter on 5 and 6 August respectively, and they raised the CENTAUR upper stage on 7 August 1997. After the Wet Dress Rehearsal on 10 September, engineers mated the spacecraft to the launch vehicle on 15 September. Officials certified the vehicle for launch on 17 September 1997. Following RP-1 propellant tanking operations, preparations continued for the launch on 5 October 1997. There were no unplanned holds during the countdown, and the ATLAS IIAS/CENTAUR lifted off Pad 36B at 2101:01.001Z on 5 October 1997. The launch was successful.[98]

ATLAS IIAS/CENTAUR (GALAXY VIII-i), AC-149, 8 December 1997

The object of the AC-149 mission was to place the GALAXY VIII-i communications spacecraft in a geosynchronous transfer orbit from Pad 36B. Astrotech Space Operations processed the spacecraft, and Lockheed Martin Astronautics integrated the payload and launched it. Hughes Space and Communications built the GALAXY VIII-i under contract to Galaxy. The spacecraft weighed 7,745 pounds, and it had an operational life expectancy of 15 years. With antennas and solar arrays deployed, the satellite was 33 feet wide and 86 feet long. The GALAXY VIII-i carried thirty-two 115-watt Ku-Band transponders. Once it was on final orbit, the satellite was expected to provide DIRECTV programming to Mexico, the Caribbean, Central America, and South America.[99]

The ATLAS IIAS/CENTAUR vehicle arrived at Cape Canaveral on 7 October 1997. Technicians erected the ATLAS and its interstage adapter on 14 October, and they raised the CENTAUR upper stage on 15 October 1997. After the Wet Dress Rehearsal on 17 November, engineers mated the spacecraft to the launch vehicle on 20 November 1997. Officials certified the vehicle for launch on 21 November 1997. Following RP-1 propellant tanking operations, preparations continued for the launch on 8 December 1997. Drifting weather balloons fouled the range about half an hour before the launch, but the incident occurred during the final built-in hold and it only lasted 17 minutes. The ATLAS IIAS/CENTAUR lifted off Pad 36B at 2352:01.501Z on 8 December 1997. The launch was successful.[100]

ATLAS IIAS/CENTAUR (INTELSAT 806), AC-151, 28 February 1998

The object of the AC-151 mission was to place the INTELSAT 806 commercial communications spacecraft in a geosynchronous transfer orbit from Pad 36B. Lockheed Martin Missiles and Space built the spacecraft, and it was processed and encapsulated at Astrotech Space Operations (ASO) in Titusville, Florida. The INTELSAT 806 payload weighed 7,772-pounds, and it was designed around the Lockheed Martin 7000-series spacecraft bus. The spacecraft had an on-orbit life expectancy of more than 10 years. From its final orbital position of 319.5 degrees east, the INTELSAT 806 would provide C-Band and Ku-Band communications services to the Americas and Europe. Lockheed Martin Astronautics built the launch vehicle, integrated the payload, and flew the mission.[101]

The ATLAS IIAS/CENTAUR chosen for the launch arrived at Cape Canaveral on 18 December 1997. Technicians erected the ATLAS and its interstage adapter on 6 January, and engineers added the CENTAUR upper stage to the launch vehicle on 7 January 1998. Technicians installed the four Solid Rocket Boosters (SRBs) assigned to the launch vehicle on 5 February 1998. After the Wet Dress Rehearsal on 13 February, engineers erected the spacecraft on 17 February 1998. Officials certified the vehicle for launch on 19 February 1998. Following RP-1 propellant tanking operations, work continued for the launch operation, which was scheduled for 27

February 1998. Thick clouds loomed overhead as the countdown began on the 27th, and instrumentation problems began to surface as well. Bermuda lost half of its radar, telemetry, and command links at 1805Z due to a power outage on the island. The Cape's High Power Command Message Encoder Verifier (CMEV) went out at 1904Z due to a bad modem card. Fortunately, technicians managed to restore the CMEV at 2010Z, just in time to see one of Antigua's telemetry antennas went down with a bad azimuth drive motor. Then, like a scene from a Hollywood movie, the clouds thinned, Bermuda's links were restored, and one of Antigua's telemetry antennas was brought online at 2129Z to replace one that was inoperable. After a 56-minute-long "NO-GO" for upper level winds cleared, the countdown proceeded without any further delay. The ATLAS IIAS/CENTAUR lifted off Pad 36B at 0021:00.831Z on 28 February 1998. The launch was a success.[102]

ATLAS IIAS/CENTAUR (INTELSAT 805), AC-153, 18 June 1998

The object of the AC-153 mission was to place the INTELSAT 805 commercial communications spacecraft in a geosynchronous transfer orbit from Pad 36A. Lockheed Martin Missiles and Space built the spacecraft, and it was processed and encapsulated at Astrotech Space Operations (ASO) in Titusville, Florida. The INTELSAT 805 payload weighed 7,772-pounds, and it was designed around the Lockheed Martin 7000-series spacecraft bus. The spacecraft had an on-orbit life expectancy of more than 10 years. The INTELSAT 805 would provide C-Band and Ku-Band communication services to the Americas and Europe. Lockheed Martin Astronautics built the launch vehicle, integrated the payload and flew the mission.[103]

The ATLAS IIAS/CENTAUR selected for the launch arrived at Cape Canaveral on 13 April 1998. Technicians erected the ATLAS and its interstage adapter on 17 and 23 April respectively, and engineers added the CENTAUR upper stage to the launch vehicle on 27 April 1998. Four Solid Rocket Boosters (SRBs) were offloaded and attached to the ATLAS booster/sustainer on 27, 28, and 29 April 1998. After the Wet Dress Rehearsal on 27 May, engineers erected the spacecraft on 8 June 1998. Officials certified the vehicle for launch on 10 June 1998. Following RP-1 propellant tanking operations, work continued for the launch on 18 June 1998. There were no unplanned holds during the countdown, and the ATLAS IIAS/CENTAUR lifted off Pad 36A at 2248:01.469Z on 18 June 1998. The launch was successful.[104]

ATLAS IIA/CENTAUR (HOTBIRD-5), AC-134, 9 October 1998

The object of the AC-134 mission was to place the HOTBIRD-5 spacecraft in a geosynchronous transfer orbit from Pad 36B. Matra Marconi Space built the spacecraft, and it was operated by

EUTELSAT to provide radio, television, and multi-media services on 20 Ku-Band channels to Europe, North Africa, and the Middle East. The spacecraft was processed and encapsulated at Astrotech Space Operations (Titusville, Florida) prior to its integration and launch by Lockheed Martin Astronautics at Cape Canaveral.[105]

The ATLAS IIA/CENTAUR selected for the mission arrived at Cape Canaveral on 10 August 1998. Technicians erected the ATLAS and its interstage adapter on 31 August 1998, and engineers added the CENTAUR upper stage to the launch vehicle on 1 September 1998. Officials completed the Wet Dress Rehearsal (WDR) successfully on 24 September, and engineers mated the spacecraft to the launch vehicle on 28 September 1998. Following launch certification and RP-1 propellant tanking operations, preparations continued for the launch on 9 October 1998. The weather was unfavorable during most of the countdown on October 9th (e.g., lightning, field mills violations, anvil cloud formations, etc.), but it improved sufficiently to allow lift-off toward the end of the launch window. The count had to be recycled at T minus 2 minutes for a booster problem, but it soon cleared. The ATLAS IIA/CENTAUR lifted off Pad 36B at 2250:01.067Z on 9 October 1998. The launch was a success.[106]

ATLAS IIAS/CENTAUR (JCSAT-6), AC-152, 16 February 1999

The object of the AC-152 mission was to place Japan's JCSAT-6 communications spacecraft in a nominal transfer orbit from Pad 36A. Hughes Space & Communications built the JSCAT-6 in El Segundo, California. The payload weighed 6,810 pounds, and it was purchased to provide commercial communications services for the Asia-Pacific region and Hawaii. The payload was processed at Astrotech's facility in Titusville, Florida. Lockheed Martin Astronautics built the launch vehicle, integrated the payload, and flew the mission.[107]

Technicians erected the ATLAS IIAS/CENTAUR booster on 10 November 1998. Four Castor IVA solid rocket motors were added to the vehicle on 16 November 1998. Following the Wet Dress Rehearsal (WDR) on 17 December 1998, engineers mated the JCSAT-6 to the launch vehicle. Unfortunately, the spacecraft developed a problem on 7 January 1999, and the JCSAT was separated from the vehicle on 10 January so a component could be replaced. Following the re-mate of the spacecraft on 21 January, officials completed the launch certification on 22 January 1999. The launch was scheduled for 1 February, but officials scrubbed it at 2136Z on 31 January 1999 after high ground winds were forecast throughout the launch window. The second launch attempt was set for 2 February, but it was scrubbed at 2210Z on 1 February so engineers could inspect a suspicious CENTAUR propellant component. The propellant component was replaced, and a new launch date of 15 February 1999 was approved. Unfortunately, the launch attempt on 15 February was scrubbed after ATLAS fuel tank pressure measurements prompted an investigation, but the final attempt on 16 February was successful. Though the countdown on the 16th was marred for four minutes by a "fouled" range and for 25 minutes by a CENTAUR anomaly, the ATLAS IIAS/CENTAUR lifted off Pad 36A at 0145:26.021Z on 16 February 1999. The launch was successful.[108]

ATLAS IIAS/CENTAUR (EUTELSAT W3), AC-154, 12 April 1999

The object of the AC-154 mission was to place the EUTELSAT W3 commercial communications spacecraft in a geosynchronous transfer orbit from Pad 36A. Aerospatiale built the spacecraft, and it was purchased to provide "high performance" telecommunications to Europe. The payload was processed and encapsulated at Astrotech Space Operations (ASO) in Titusville, Florida. Lockheed Martin Astronautics built the launch vehicle, integrated the payload, and flew the mission.[109]

Technicians erected the ATLAS IIAS/CENTAUR booster on 24 February 1999. Four Castor IVA solid rocket motors were added to the vehicle on 26 February, and officials completed the Wet Dress Rehearsal (WDR) on 30 March 1999. Following spacecraft mate on 3 April, a Composite Electrical Readiness Test (CERT) was completed on 5 April 1999. Preparations continued for a launch on 12 April 1999. Unlike the JCSAT mission, the EUTELSAT W3 only required one launch attempt to send it on its way. The countdown on 12 April went fairly smoothly, though a couple of rough spots were noted. One of the range's radars went Non-Mission Capable (NMC) due to a transmitter problem at 1830Z. Technicians discovered a failed tube and replaced it, and the radar was Fully Mission Capable (FMC) at 1900Z. Station 1 Command was Partially Mission Capable (PMC) at 1845Z due to problems with the Antenna Designate Unit (ADU). Troubleshooters investigated the problem, and Station 1 Command returned to Fully Mission Capable (FMC) status at 1948Z. Those problems aside, the countdown proceeded without any unscheduled holds, and the ATLAS IIAS/CENTAUR lifted of Pad 36A at 2250:01.352Z on 12 April 1999. The launch was successful.[110]

ATLAS IIAS/CENTAUR (ECHOSTAR V, MCI-1), AC-155, 23 September 1999

The object of the AC-155 mission was to place the ECHOSTAR V (MCI-1) commercial communications satellite in a 28,455-mile-high super-synchronous transfer orbit from Pad 36A. Space Systems/Loral manufactured the spacecraft, which weighed 7,925 pounds. Astrotech Space Operations processed the payload in Titusville, Florida. Lockheed Martin Astronautics provided the ATLAS IIAS/CENTAUR vehicle, integrated the payload, and launched it into space. The DISH (Digital Sky Highway) Network purchased the spacecraft to become part of the network's six-satellite constellation, which received its first satellite in 1995. From its orbit (positioned 110 degrees west longitude), the ECHOSTAR V would provide high definition digital communications services to the continental United States, Hawaii, Alaska, and Puerto Rico.[111]

Technicians erected the ATLAS IIAS/CENTAUR booster on 5 August, and officials completed the Wet Dress Rehearsal (WDR) on 27 August 1999. Following spacecraft mate on 31 August 1999, Lockheed Martin cleared the vehicle for its first launch attempt on 10 September 1999.[112] Unfortunately, a severe thunderstorm on 6 September caused lightning damage on Pad 36A's Umbilical Tower, and water penetrated the ATLAS' B-1 Pod. The pad dried out, and engineers retested flight/ground hardware successfully on 9 September 1999. Officials selected a new launch date of 13 September 1999, but Lockheed Martin decided to postpone the launch until the corporation's engineers could analyze an ATLAS Remote Control Unit (RCU) that failed factory testing on 10 September 1999. Investigators determined the failed component was an isolated case, but the threat of Hurricane FLOYD intervened a few days later, so the launch was rescheduled for 22 September 1999. Tropical Storm HARVEY threatened the launch with a 100 percent chance of rain on the 22nd, so officials postponed the launch until 23 September 1999. Finally, on 23 September 1999, the countdown got underway. Though two unauthorized boats and two unauthorized ships in the sea surveillance area fouled the range for about one hour and 17 minutes,[113] the ATLAS IIAS/CENTAUR lifted off Pad 36A at 0602:01.677Z on 23 September 1999. The vehicle placed the ECHOSTAR V in the proper super-synchronous transfer orbit as planned.[114]

ATLAS IIAS/CENTAUR (HISPASAT 1-C), AC-158, 3 February 2000

The object of the AC-158 mission was to place the HISPASAT 1-C commercial communications spacecraft in a geosynchronous transfer orbit from Pad 36B. Alcatel Space Systems of France manufactured the spacecraft, and the HISPASAT 1-C was purchased to provide digital television broadcast services in Spanish to customers in Spain and Latin America. The payload was processed and encapsulated at Astrotech Space Operations (ASO) in Titusville, Florida. Lockheed Martin Astronautics built the launch vehicle, integrated the payload, and flew the mission. According to the mission scenario, the ATLAS IIAS/CENTAUR would insert the spacecraft into a parking orbit. With the CENTAUR stage still attached, the spacecraft would coast in the parking orbit until the CENTAUR fired a second time to insert the HISPASAT 1-C in the proper transfer orbit. The satellite would reach its final geostationary orbit using its onboard propulsion resources.[115]

Technicians erected the ATLAS IIAS/CENTAUR booster on Pad 36B on 15 December 1999. Four Castor IVA solid rocket motors were added to the vehicle on 4 January 2000, and officials completed the WDR on 18 January 2000. The spacecraft mating was scheduled for 25 January, but problems with a bridge crane and excessive winds on Pad 36B delayed that milestone for two days.[116] Following the spacecraft mate on 27 January, a Composite Electrical Readiness Test (CERT)) was completed on 28 January 2000. Preparations for the launch on 3 February were hampered with bad weather during most of the countdown, and the T minus 5 minutes hold continued for 20 minutes due to weather and a battery problem. Nevertheless, the weather improved sufficiently to permit the Weather Officer to declare the range 'Green' for launch at 2320Z. The ATLAS IIAS/CENTAUR lifted of Pad 36B at 2330:00.752Z on 3 February 2000. The launch was successful.[117]

ATLAS IIAS/CENTAUR (ECHOSTAR VI), AC-161, 14 July 2000

The object of the AC-161 mission was to place the ECHOSTAR VI commercial communications satellite in a geosynchronous transfer orbit from Pad 36B. According to the flight scenario, the ATLAS IIAS/CENTAUR would place the payload in a low Earth parking orbit initially. Following a brief coasting period, the CENTAUR upper stage would fire again (e.g., 2nd Burn) to place the ECHOSTAR VI in a geosynchronous transfer orbit. The spacecraft would separate from the CENTAUR after that, and the ECHOSTAR VI would use its own propulsion system to climb to a geostationary orbit. Space Systems/Loral built the spacecraft, and it was purchased by the EchoStar Communications Corporation to provide direct-to-home television broadcast services to EchoStar's customers in North America. The spacecraft was processed and encapsulated at Astrotech's space operations facility in Titusville, Florida. Lockheed Martin Astronautics was responsible for integrating the payload and launching the ATLAS IIAS/CENTAUR vehicle.[118]

Technicians erected the ATLAS IIAS/CENTAUR booster on 7 June 2000. The vehicle's four Solid Rocket Motors (SRMs) should have been mated on 15 June, but electrical problems with the Torus crane delayed that operation until 22 June 2000.[119] Officials completed the Simulated Flight Test and Wet Dress Rehearsal on schedule (e.g., 20 and 26 June 2000 respectively), but the ECHOSTAR VI/launch vehicle mating operation was delayed from 30 June to 6 July due to a change in the launch date for the TDRS-H mission. Officials completed the Composite Electrical Readiness Test (CERT) on 7 July, and preparations continued for the launch on 14 July 2000. There were no unplanned holds during the countdown on 13/14 July, but a radar and a range camera site were Non-Mission Capable (NMC) briefly when they experienced a power bump at 2331Z on 13 July. They were both restored to Fully Mission Capable (FMC) status within one hour. One of Antigua's circuits experienced short dropouts at 0455Z on 14 July. The cause for the dropouts could not be determined without taking the circuit down, but that would have rendered the range 'Red' very late in the count. Since the dropouts cleared and visual inspections revealed no further problems, CSR decided to commit the line about 15 minutes later without taking the circuit down. The Range Control Officer received the final clear to launch at 0512Z, and the vehicle lifted off Pad 36B at 0521:01.431Z on 14 July 2000. The launch was successful.[120]

ATLAS IIAS/CENTAUR (ICO-A1 [F2]), AC-156, 19 June 2001

The object of the AC-156 mission was to place the New ICO-A1 (F2) communications spacecraft into a medium-Earth orbit (MEO) from Pad 36B. An ATLAS IIAS/CENTAUR would be used to boost the spacecraft into orbit along a flight azimuth of 53.2 degrees. The satellite's ultimate destination was a 6,456-mile-high orbit in one of two orthogonal planes inclined 45 degrees to the equator.[121] Astrotech processed the spacecraft at its facility in Titusville, Florida. Lockheed Martin Astronautics provided the ATLAS IIAS/CENTAUR vehicle, integrated the payload, and launched the entire assembly into space.[122]

Technicians erected the ATLAS IIAS/CENTAUR booster on 23 April 2001. Following the Wet Dress Rehearsal (WDR) on 18 May, technicians connected four Castor IVA solid rockets to the vehicle on 24 May 2001. Engineers mated the ICO A-1 (F2) spacecraft to the vehicle on 9 June 2001. Officials completed the Composite Electrical Readiness Test (CERT) on 11 June, and the countdown proceeded as scheduled on 19 June 2001. Though there were a number of minor data line, optics, and weather instrumentation problems during the countdown, the range remained 'Green' throughout the count. The ATLAS IIAS/CENTAUR lifted off Pad 36B at 0441:01.770Z on 19 June 2001. The launch was successful. With its first A-1 spacecraft on orbit, New ICO could begin testing the hardware associated with its 12-satellite constellation. The network was slated to begin operations in the 2003-2004 period.[123]

ATLAS IIAS/CENTAUR (HISPASAT-1D), AC-159, 18 September 2002

The object of the AC-159 mission was to place the Spanish HISPASAT-1D communications spacecraft in a geosynchronous transfer orbit from Pad 36A. Alcatel Space Systems built the 7,000-pound spacecraft (based on the company's Spacebus 3000B2), and the satellite was scheduled to replace the HISPASAT 1A and 1B spacecraft launched in 1992 and 1993. The HISPASAT-1D was equipped with 28 Ku-band transponders, and it was designed to provide direct-to-home television services in the Americas and Europe. The spacecraft and its ground support equipment were transported to Astrotech facilities in Titusville, Florida, for pre-launch processing. Lockheed Martin Astronautics built the launch vehicle, integrated the payload, and flew the mission.[124]

The launch was scheduled for 18 September 2002, and the launch team accomplished their pre-launch preparations according to plan. Technicians erected the ATLAS IIAS booster on 15 July, and they finished installing the Castor IVA solid rocket motors on 17 July 2002. Officials completed the Wet Dress Rehearsal on 29 August, and engineers mated the payload to the launch vehicle on 9 September 2002. The Composite Electrical Readiness Test was completed successfully on 11 September 2002, and the countdown went smoothly one week later. There were no unplanned holds during the count, and the ATLAS IIAS/CENTAUR lifted off Pad 36A at 2204:00.615Z on 18 September 2002. The launch was successful.[125]

ATLAS IIAS/CENTAUR (AMC-10), AC-165, 5 February 2004

The object of the AC-165 mission was to place SES AMERICOM's AMC-10 commercial communications spacecraft in a proper transfer orbit so it could replace the company's SATCOM C3 spacecraft. The AMC-10 featured double and triple backups of mission-critical systems to ensure delivery of service throughout the 15-year operational life of the satellite. Together with AMC-11, AMC-10 was designed to provide advanced C-Band digital communications to SES AMERICOM's regional and national customers across the United States (e.g., Viacom, C-SPAN, Lifetime Entertainment, The Weather Channel, and others). Lockheed Martin built the AMC-10 and AMC-11 spacecraft. Lockheed Martin Astronautics built the launch vehicle, integrated the payload, and flew the mission.[126]

Technicians erected the ATLAS IIAS/CENTAUR on Pad 36A on 10 November 2003. The AMC-10 arrived on 19 December 2003. Officials conducted the Simulation Flight and Wet Dress Rehearsal on 9 and 20 January 2004 respectively, and engineers mated the AMC-10 to the launch vehicle on 28 January 2004. The spacecraft mate was delayed from 27 January to 28 January due to weather constraints, and the Combined Electrical Readiness Test was delayed until 30 January 2004 as a consequence of the spacecraft mating delay. Neither delay impacted the launch, which officials had rescheduled from 27 February 2004 to 5 February 2004 back in late September 2003. Preparation continued for the countdown on the 5th as planned.[127]

The countdown went well. There was one unplanned hold in the count at T minus 65 minutes (e.g., to deal with a booster helium charge problem), but the ATLAS IIAS/CENTAUR lifted off

Pad 36A at 2346:02.651Z on 5 February 2004. The payload was inserted into the proper transfer orbit at 0014:14.000Z on 6 February 2004.[128]

ATLAS IIAS/CENTAUR (SUPERBIRD-6), AC-163, 16 April 2004

The object of the AC-163 mission was to place the SUPERBIRD-6 commercial communications spacecraft into a proper geosynchronous transfer orbit. Boeing Satellite Systems (BSS) built the SUPERBIRD-6 based on the BSS-601 satellite platform. It was the third in a series of Boeing satellites built for the Space Communications Corporation (SCC), a Tokyo-based firm. The SUPERBIRD-6 weighed 6,825 pounds, and it carried 23 Ku-Band transponders and four Ka-Band transponders to provide telecommunications for users in Japan. Astrotech Space Operations Company processed the SUPERBIRD-6 in Titusville, Florida. Lockheed Martin Space Systems Company processed the ATLAS IIAS/CENTAUR launch vehicle, integrated the payload, and launched it from Pad 36A at Cape Canaveral.[129]

Technicians erected the ATLAS IIAS/CENTAUR on 20 February 2004. Officials conducted the Simulation Flight and Wet Dress Rehearsal on 24 and 31 March 2004 respectively, and engineers mated the SUPERBIRD-6 to the launch vehicle on 7 April 2004. Officials completed the Combined Electrical Readiness Test on 8 April 2004. Preparations continued for the scheduled launch on 16 April 2004.[130]

The countdown on 15/16 April 2004 went well. There were no unplanned holds, and the ATLAS IIAS/CENTAUR lifted off Pad 36A at 0045:00.306Z on 16 April 2004. The SUPERBIRD-6 was inserted into its proper transfer orbit about 30 minutes after lift-off (e.g., 0115:05Z).[131]

ATLAS IIAS/CENTAUR (AMC-11) AC-166, 19 May 2004

The object of the AC-166 mission was to place the AMC-11 communications spacecraft in a proper transfer orbit. Lockheed Martin built the AMC-11 for SES AMERICOM to relay cable television programming to customers in the United States. Featuring double and triple backups to its mission-critical systems, the AMC-11 had a projected operational life expectancy of 15 years. The AMC-11 was destined to replace the SATCOM C4 spacecraft in SES AMERICOM's satellite constellation. Lockheed Martin Space Systems Company processed the ATLAS IIAS/CENTAUR launch vehicle, integrated the payload, and launched it from Pad 36B at Cape Canaveral.[132]

On 3 February 2004, officials rescheduled the AMC-11 launch from 17 May to 19 May 2004 to de-conflict upcoming launch vehicle schedules. Processing milestones slipped somewhat in late

April, but their accomplishment did not impact the launch schedule. Official completed the Simulated Flight and the Wet Dress Rehearsal on 27 April and 4 May 2004 respectively. Engineers mated the AMC-11 to the launch vehicle on 11 May 2004. Following a successful Combined Electrical Readiness Test on 13 May, processing continued for the launch on 19 May 2004.[133]

The countdown on the 19th went well. There were no unplanned holds, and the ATLAS IIAS/CENTAUR lifted off Pad 36B at 2222:00.513Z on 19 May 2004. According to the 45th Space Wing Command Post, the spacecraft was inserted into the proper transfer orbit approximately 28 minutes later.[134]

ATLAS I/CENTAUR and ATLAS II/CENTAUR Civil Space Operations

ATLAS I/CENTAUR (GOES I), AC-73, 13 April 1994

The object of the AC-73 mission was to place the GOES I payload into a highly elliptical transfer orbit. The GOES I was the first in a new series of weather satellites built by Space Systems/Loral for the National Oceanic and Atmospheric Administration (NOAA).[135] The satellite introduced two new features: 1) a flexible scan capability to let meteorologists concentrate their data-gathering on local weather trouble spots and 2) simultaneous imaging and sounding, which allowed multiple measurements of weather phenomena to increase the accuracy of weather forecasts. An ATLAS I/CENTAUR was chosen as the launch vehicle for the mission.[136]

Technicians erected the vehicle's booster/sustainer on Pad 36B on 24 January 1994. They raised the interstage adapter on 25 January, and engineers added the CENTAUR upper stage on 26 January 1994. Officials completed a flight simulation on 3 March, and they accomplished the Wet Dress Rehearsal on 10 March 1994. Following spacecraft/vehicle mate toward the end of March, technicians and engineers prepared the vehicle for launch on 13 April 1994. Apart from a booster hydraulic problem that caused a 10-minute delay in the T minus 105-minute built-in hold, the countdown went smoothly. The vehicle lifted off at 0604:02.317Z on 13 April 1994. Range support was highly satisfactory, and the mission was successful.[137]

ATLAS I/CENTAUR (GOES-J), AC-77, 23 May 95

The object of the AC-77 mission was to place the GOES-J weather satellite into a proper geosynchronous transfer orbit. The GOES-J was the second in a series of satellites built by Space Systems/Loral for the National Oceanic and Atmospheric Administration (NOAA). The GOES-J was designed to provide pictures of approximately one-fourth of Earth's surface at 30-minute intervals, 24 hours a day. Scientists used the pictures to form more precise weather predictions and to examine ocean currents, river water levels and other weather-related phenomena. The GOES program also supported the cooperative international Search and Rescue (SAR) program.[138]

The ATLAS I/CENTAUR vehicle was received at Cape Canaveral on 7 April 1995. Technicians erected the ATLAS and its interstage adapter on Pad 36B on 10 April, and engineers installed the CENTAUR upper stage on 12 April 1995. Following the launch vehicle simulated flight test on 26 April, officials conducted the Wet Dress Rehearsal on 3 May 1995. The GOES-J was encapsulated on 5 May, and engineers mated the spacecraft to the launch vehicle on 6 May 1995. After the launch certification was completed on 9 May, preparations continued for a launch attempt on 19 May 1995. Unfortunately, a faulty cell in an ATLAS battery forced officials to scrub the first launch attempt on 19 May. Weather conditions on 20 May were not good, and authorities eventually scrubbed the mission for local thunderstorms and excessive winds at 0541Z on 20 May 1995. The countdown for the third launch attempt on 23 May went well. There was a ten-minute unscheduled hold to resolve a false flight termination system continuity indication, but the rest of the countdown went smoothly. The ATLAS I/CENTAUR lifted off Pad 36B at 0552:01.893Z on 23 May 1995. The launch was successful.[139]

ATLAS IIAS/CENTAUR (SOHO), AC-121, 2 December 1995

The object of the AC-121 mission was to launch the Solar and Heliospheric Observatory (SOHO), a spacecraft sponsored by NASA and the European Space Agency (ESA). The SOHO was designed to study the Sun's internal structure, its outer atmosphere, and the origin of the solar wind. Matra Marconi Space built the spacecraft at the company's facility in Velizy, France. The SOHO consisted of two modules: 1) a service module weighing 2,734 pounds and 2) a payload module weighing 1,344 pounds. An ATLAS IIAS/CENTAUR was used to boost the SOHO on its way to a permanent vantage point 1,500,000 kilometers ahead of the Earth in a halo orbit around the L1 Lagrangian point. Scientists designed the SOHO around nine European and three American experiments, and the spacecraft was expected to observe the Sun for at least two years.[140]

The ATLAS IIAS/CENTAUR vehicle was delivered to Cape Canaveral on 15 September 1995. Technicians erected the ATLAS stage and interstage adapter on Pad 36B on 28 September, and engineers raised the CENTAUR upper stage on 29 September 1995. Following the launch vehicle simulated flight test on 24 October, officials completed the Wet Dress Rehearsal on 31

October 1995. Technicians installed the last of the vehicle's solid rocket motors on 7 November, and engineers mated the SOHO to the launch vehicle on 11 November 1995. The launch certification was completed on 14 November, and preparations continued for the first launch attempt on 23 November 1995. Unfortunately, a liquid oxygen regulator valve malfunctioned during the countdown, and officials scrubbed the launch at 0343Z on 23 November. The launch was delayed until 2 December 1995 because the valve was made of the wrong material, and it had to be flown back to Lockheed Martin for replacement. The countdown on 2 December went well. There was a 34-minute delay in the countdown due to a battery voltage spike in the vehicle's hydrazine heater, but the count was recycled and lift-off was adjusted to 0808Z. The ATLAS IIAS/CENTAUR lifted off Pad 36B at 0808:00.863Z on 2 December 1995. The launch was successful.[141]

ATLAS I/CENTAUR (GOES-K), AC-079, 25 April 1997

The object of the AC-079 mission was to place the GOES-K weather satellite into a geostationary transfer orbit from Pad 36B. The mission featured the last launch of an ATLAS I vehicle. The GOES-K was the third in a series of advanced weather satellites managed by the National Oceanic and Atmospheric Administration (NOAA). The satellite was designed to provide pictures of one-quarter of Earth's surface at 15-minute intervals, 24 hours a day. The GOES-K provided evidence of ocean currents, river water levels, and distinct weather occurrences that helped scientists predict weather patterns. The spacecraft also supplemented cooperative international Search and Rescue (SAR) efforts. Space Systems/Loral built the spacecraft.[142]

The ATLAS I/CENTAUR vehicle arrived at Cape Canaveral on 25 March 1997. Technicians erected the ATLAS and its interstage adapter on 10 and 11 March respectively, and engineers raised the CENTAUR upper stage on 12 March 1997. After the Wet Dress Rehearsal on 8 April, engineers mated the spacecraft to the launch vehicle on 11 April. Officials certified the vehicle for launch on 15 April 1997. Following RP-1 propellant tanking operations, preparations continued for the launch on 25 April 1997. There were no unplanned holds during the countdown on April 25th, and the ATLAS I/CENTAUR lifted off Pad 36B at 0549:01.443Z on 25 April 1997. The launch was successful.[143]

ATLAS IIA/CENTAUR (GOES-L), AC-137, 3 May 2000

The object of the AC-137 mission was to place the latest in a series of Geostationary Operational Environmental Satellites (GOES-L) in a transfer orbit from Pad 36A. The GOES-L was manufactured Space Systems/Loral, and the spacecraft weighed 4,890 pounds.[144] Lockheed Martin Astronautics provided the ATLAS IIA/CENTAUR vehicle, integrated the payload, and launched the entire assembly into space. According to the mission scenario, the ATLAS IIA/CENTAUR would place the payload in a low Earth parking orbit. After a brief coasting phase, the CENTAUR would fire a second time to place the GOES-L in a geosynchronous transfer orbit. Following that second burn, the CENTAUR would separate from the spacecraft, and the GOES-L would insert itself into final orbit under its own power. The spacecraft was added to the GOES constellation as an on-orbit spare at 75 degrees west longitude. It was destined to replace GOES-8 as GOES-East later in 2000.[145]

The mission was originally scheduled for launch on 15 May 1999, but it was delayed almost a year. Initially, technicians erected the ATLAS IIA/CENTAUR booster on Pad 36A on 17 April 1999. Following a Simulated Flight Test on 27 April, officials completed a Wet Dress Rehearsal (WDR) on 4 May 1999. Engineers mated the spacecraft to the launch vehicle on 6 May, and the Composite Electrical Readiness Test (CERT)) was completed on 7 May 1999. Lockheed Martin cleared the vehicle for its first launch attempt, but NASA requested a delay to review data from the DELTA III launch mishap on May 5th.[146] The launch was pushed to 23 May, then 30 May, then 11 June, then 16 July 1999. In the meantime, the spacecraft's solar panel actuator mechanisms were in danger of oxidizing due to the extended period the payload was on the launch pad. To avoid that eventuality, engineers demated the GOES-L from the ATLAS IIA/CENTAUR on 15 June 1999, and the payload was transported to Astrotech's processing facility in Titusville, Florida. Engineers and technicians de-stacked the ATLAS IIA/CENTAUR by 17 July 1999.[147]

The second attempt to assemble and launch the GOES-L mission fared better in 2000. Technicians erected the ATLAS IIA/CENTAUR on Pad 36A on 27 March 2000. Due to a conflict with the AC-201 schedule, officials completed the Wet Dress Rehearsal and the Simulated Flight Test for the GOES-L on 11 April 2000 instead of 7 April. Fortunately, the delay had no effect on the launch date. Engineers mated the spacecraft to the launch vehicle on 21 April, and the Composite Electrical Readiness Test (CERT) was accomplished on 24 April 2000. Though the countdown on 3 May was held at T minus 105 minutes for 40 minutes due to "problems at [the] pad," the countdown and flight were nominal. The ATLAS IIA/CENTAUR lifted off Pad 36A at 0707:02.866Z on 3 May 2000. The spacecraft separated from the vehicle at an altitude of 22,876 nautical miles, and it entered super-synchronous transfer orbit. The launch was successful.[148]

ATLAS IIA/CENTAUR (TDRS-H), AC-139, 30 June 2000

The object of the AC-139 mission was to place NASA's Tracking and Data Relay Satellite (TDRS-H) in a sub-synchronous transfer orbit from Pad 36A. According to the flight scenario, the ATLAS IIA/CENTAUR would place the payload in a low Earth parking orbit initially. Following a brief coasting period, the CENTAUR upper stage would fire again (e.g., 2nd Burn) to place the TDRS-H in a sub-synchronous transfer orbit. The spacecraft would separate from the CENTAUR after that, and the TDRS-H would use its onboard propulsion system to climb to a geosynchronous orbital position at 150 degrees West Longitude. Following on-orbit testing, the spacecraft would be moved to its operational position at 171 degrees West Longitude. The 7,000-pound spacecraft was based on the Hughes HS601 satellite bus, which was manufactured by the Hughes Space and Communications Company. The TDRS-H was the first of three new satellites (TDRS-H, I, and J) designed to upgrade critical U.S. relay satellite capabilities for NASA, the U.S. intelligence community, the European Space Agency, the Japanese Space Agency, and commercial customers. All together, the TDRS-H, I, and J spacecraft and their launch vehicles were worth about $830 million. The Goddard Space Flight Center (NASA) managed the program.[149]

Technicians erected the ATLAS IIA/CENTAUR booster on Pad 36A on 22 May 2000. Lockheed Martin slipped the Simulated Flight Test from 6 June to 7 June 2000 to make the most of contractor resources, but this change did not impact any other milestones. Officials completed the Wet Dress Rehearsal on 14 June as planned. NASA requested a one-day delay to verify ground equipment readiness on 15 June, so the spacecraft was mated to the vehicle on 17 June 2000 instead of 16 June. The Composite Electrical Readiness Test (CERT)) was accomplished on 19 June, but NASA asked to delay the AC-139 launch from 29 June to 30 June 2000 to give NASA time to evaluate findings on a recent RL-10 engine investigation.[150] The countdown on 30 June went fairly smoothly, but it was delayed after a boat entered the launch danger area at 1230Z. The range went 'Red' due to that intrusion, and the T minus 5 minutes hold was extended at 1233Z. A new T-0 of 1256Z was coordinated, and the range became 'Green' at 1243Z after the boat cleared the one-boat contour. The count was picked up at 1251Z, and the ATLAS IIA/CENTAUR lifted off Pad 36A at 1256:00.662Z on 30 June 2000. The launch was successful, and the TDRS-H was placed in the proper orbit.[151]

ATLAS IIA/CENTAUR (GOES-M), AC-142, 23 Jul 2001

The object of the AC-142 mission was to place the Geostationary Operational Environmental Satellite M (GOES-M) spacecraft in a super-synchronous transfer orbit[152] from Pad 36A. The spacecraft weighed approximately 5,000 pounds, and it provided data for weather forecasts and Search and Rescue (SAR) operations all over the world. Space Systems/Loral manufactured the $240 million spacecraft for the National Oceanic and Atmospheric Administration (NOAA). NOAA planned to use GOES-M as a critical spare initially. As such, the spacecraft would serve as a backup for: 1) the GOES 8, which covered America's East and Gulf coasts, and 2) the GOES 10, which covered America's West Coast and the Eastern Pacific Ocean. Lockheed Martin Astronautics built the launch vehicle, integrated the payload, and flew the mission.[153]

Technicians erected the ATLAS IIA/CENTAUR booster on Pad 36A on 31 May 2001. Officials completed the Wet Dress Rehearsal on 29 June, and contractor personnel mated the spacecraft to the vehicle on 6 July 2001. A module in the CENTAUR's Remote Control Unit (RCU) failed during testing on 9 July, and the launch had to be delayed from 15 July to 22 July 2001 while the module was replaced and checked out. Following the completion of a successful Composite Electrical Readiness Test on 16 July, the contractor completed final preparations for launch. Two launch attempts were required to complete the mission. There were no major instrumentation problems during the first countdown on 22 July, but the user scrubbed the first launch attempt after lightning struck Pad 36B at 0156Z on the 22nd.[154] Officials noted only one major instrumentation anomaly during the countdown on 23 July. A radar went Non-Mission Capable (NMC) at 0635Z for a control power supply failure, but the site was declared Fully Mission Capable (FMC) at 0655Z after a resistor was replaced to bring the control power supply back online. In the meantime, a boat intruded into the launch danger area. Officials declared a "No-Go" for safety from 0650Z until the boat cleared the area at 0715Z. Without much further ado, the ATLAS IIA/CENTAUR lifted off Pad 36A at 0723:01.299Z on 23 July 2001. The launch was successful.[155]

ATLAS IIA/CENTAUR (TDRS-I), AC-143, 8 March 2002

The object of the AC-143 mission was to place NASA's Tracking and Data Relay Satellite (TDRS-I) in a sub-synchronous transfer orbit from Pad 36A. The TDRS-I was the second of three new 7,000-pound satellites (TDRS-H, I, and J) designed to upgrade critical U.S. relay satellite capabilities for NASA, the U.S. intelligence community, the European Space Agency, the Japanese Space Agency, and commercial customers. Successful introduction of the new satellites would extend the TDRS System's service life through 2017. The Goddard Space Flight Center (NASA) managed the program.[156]

In accordance with accepted procedures, the TDRS-I spacecraft and its ground support equipment were shipped by truck and air transport to the Kennedy Space Center's Spacecraft

Assemble Encapsulation Facility #2 for pre-launch processing. Lockheed Martin Astronautics provided the ATLAS IIA/CENTAUR vehicle, integrated the payload, and launched the entire assembly into space. According to the flight scenario, the ATLAS IIA/CENTAUR selected for the mission would roll into a flight azimuth of 95.4 degrees. The CENTAUR's first burn would place the upper stage and payload in a low Earth parking orbit initially. Following a brief coasting period, the CENTAUR's second burn would inject the TDRS-I into the planned sub-synchronous transfer orbit. Following on-orbit testing, the spacecraft would be moved into its operational position.[157]

Technicians erected the ATLAS IIA/CENTAUR booster on Pad 36A on 17 September 2001. Though officials planned to conduct the Wet Dress Rehearsal on 16 October 2001, a problem with the TDRS-H's communications relay antennas surfaced in December 2001.[158] The problem required extensive troubleshooting from December 2002 through mid-January 2003 before officials permitted the shipment of the TDRS-I to KSC. In the meantime, officials rescheduled the Wet Dress Rehearsal for 9 January 2002. The launch slipped to 6 February before sliding to 8 March 2002. Ultimately, the TDRS-I arrived on 4 February, and it was mated to the vehicle on 26 February 2002. Officials completed the Composite Electrical Readiness Test on 28 February, and the countdown proceeded as rescheduled on 8 March 2002.[159]

Though the launch was ultimately successful, there were two incidents during the countdown that deserve special mention. At 1719Z, a radar was declared Partially Mission Capable (PMC) when power was cut off in a building for an air-conditioning problem. Since the building contained a modem that passed timing to the radar, the loss of power 'froze' the L-clock and generated the PMC condition. The Range Control Officer (RCO) contacted Patrick's civil engineers, and they restored power so the radar could support the mission. The radar was declared Fully Mission Capable (FMC) at 1815Z. The other incident involved the detection and interception of unidentified aircraft flying due east of Patrick AFB at T minus 5 minutes. At T minus 4 minutes, the aircraft began heading north at about 120 knots per hour, and the Safety Control Officer (SCO) dispatched an Air Force Search and Rescue (SAR) helicopter to "intercept [the aircraft] and render any assistance" *after* the ATLAS IIA/CENTAUR "cleared the area." At T plus 4 minutes, the SAR helicopter reported intercepting a flight of three U.S. Navy Seahawk helicopters. The SAR helicopter joined the Navy formation and established radio contact. Officials determined later that the Navy helicopters had a malfunctioning transponder and the aircrews had not been briefed on the ATLAS IIA/CENTAUR launch.[160]

In addition to those incidents, there was a 20-minute-long delay in the countdown due to the user's ground systems. During that delay, officials declared the range 'Red' for about six minutes to allow SAR helicopters time to verify the launch danger area was clear of targets. The scheduled T-0 slipped from 2239Z to 2259Z, and the ATLAS IIA/CENTAUR lifted off Pad 36A at 2259:00.935Z on 8 March 2002. The launch was successful.[161]

ATLAS IIA/CENTAUR (TDRS-J), AC-144, 5 December 2002

The object of the AC-144 mission was to place NASA's Tracking and Data Relay Satellite (TDRS-J) in a sub-synchronous transfer orbit from Pad 36A. The TDRS-J was the third of three new satellites (TDRS-H, I, and J) designed to upgrade critical U.S. relay satellite capabilities for NASA, the U.S. intelligence community, the European Space Agency, the Japanese Space Agency, and commercial customers. The Goddard Space Flight Center (NASA) managed the program.[162]

The TDRS-J spacecraft and its ground support equipment were shipped by truck and air transport to the Kennedy Space Center's Spacecraft Assemble Encapsulation Facility #2 for pre-launch processing. Lockheed Martin Astronautics provided the ATLAS IIA/CENTAUR launch vehicle, integrated the payload, and launched the mission. On a flight heading of 95.4 degrees, the CENTAUR would fire to place the upper stage and payload in a low Earth parking orbit initially. Following a brief coasting period, the CENTAUR's second burn would inject the TDRS-J into the planned sub-synchronous transfer orbit. Following on-orbit testing, the spacecraft would be moved into its operational position.[163]

Technicians erected the ATLAS IIA/CENTAUR booster on Pad 36A on 9 October 2002. Officials conducted the Wet Dress Rehearsal on 6 November, and engineers mated the spacecraft to the launch vehicle on 22 November 2002. The Composite Electrical Readiness Test was completed successfully on 26 November 2002. There were no unscheduled holds during the countdown on 4/5 December 2002, and the ATLAS IIA/CENTAUR lifted off Pad 36A without incident at 0242:01.181Z on 5 December 2002. The launch was successful.[164]

ATLAS IIIA/CENTAUR and ATLAS IIIB/CENTAUR Overview

The Program Requirements Document (PRD) for commercial ATLAS IIIA/CENTAUR launches (PRD 5400) was published on 20 April 1998 as an outgrowth of the ATLAS IIAS Program Introduction (dated 15 February 1991). No substantial modifications of the launch pads on Complex 36 were required to accommodate the new vehicle, and Lockheed Martin's ceremony for the arrival of first ATLAS IIIA was held at Pad 36B on 9 March 1999. Operations were underway shortly thereafter. Following the first launch of Lockheed Martin's ATLAS

IIIA/CENTAUR from Pad 36B on 24 May 2000, the corporation capped that success with the first ATLAS IIIB/CENTAUR launch from Pad 36B on 21 February 2002.[165]

The ATLAS IIIA capitalized on the pressure-stabilized design of the ATLAS II, but it also introduced a new single-stage engine: the Russian RD-180. The RD-180 was a throttleable, two-chamber version of NPO Energomash's reliable old RD-170. The ATLAS IIIA and ATLAS IIIB both used the new engine, which burned RP-1 (kerosene) and produced 860,200 pounds of thrust at sea level. The ATLAS III design (both variants) eliminated the stage-and-a-half booster/sustainer used on the ATLAS II. They employed a single 10 x 95.1-foot booster instead. Other differences between the ATLAS III series and the ATLAS II series included: 1) no Castor IVA Solid Rocket Motors (SRMs), 2) replacement of two Pratt & Whitney RL10A-4 engines in the CENTAUR upper stage with either one or two Pratt & Whitney RL10A-4-1 engine(s),[166] 3) replacement of the ATLAS II's hydraulic Thrust Vector Control (TVC) system with electromechanical TVC actuators, and 4) lengthening of the ATLAS II liquid oxygen tank, engine fairing, interstage adapter, and payload fairings. With the large payload fairing in place, the ATLAS IIIA/CENTAUR stood 170.2 feet tall. The ATLAS IIIB/CENTAUR stood 174.2 feet tall in similar payload configuration. The ATLAS IIIB/CENTAUR could lift payloads weighing as much as 9,920 pounds to geosynchronous transfer orbit (GTO).[167]

In a typical ATLAS III/CENTAUR flight scenario (either "A" or "B" variant), the launch vehicle's RD-180 ignited about one or two seconds before lift-off. Once its fuel was depleted (e.g., about three minutes into the flight), the RD-180 shut down and separated from the CENTAUR upper stage and payload. The CENTAUR fired about 11 seconds after separation, and the first CENTAUR burn injected the upper stage and spacecraft into a parking orbit. The payload fairing was jettisoned during the CENTAUR's first burn. Following a coasting period of roughly nine minutes, the CENTAUR restarted and fired for about three minutes to place the payload in a geosynchronous transfer orbit. The spacecraft separated from the upper stage some minutes later, and it used its onboard propulsion system to achieve final orbit.[168]

Range instrumentation for a typical ATLAS III/CENTAUR mission included radars on the Cape, Merritt Island, Patrick AFB, Antigua, and Jonathan Dickinson Missile Tracking Annex. Command transmitter/exciters at the Cape, Jonathan Dickinson, and Antigua provided command and control. HH-60 helicopters operated out of Patrick AFB to provide safety sea surveillance, and an SCE Learjet provided weather reconnaissance. Telemetry units on Merritt Island, Jonathan Dickinson, and Antigua supported the flights, and so did the Tracking and Data Relay Satellite System (TDRSS). Optical support included Patrick's Fixed Intercept Ground Optical Recorder (IGOR) or the Cocoa Beach Distant Object Attitude Measuring System (DOAMS), as well as cameras at other range optical sites. Patrick's WSR-74C weather surveillance radar, the WSR-88D radar in Melbourne, and a whole host of weather instrumentation systems at Cape Canaveral supported each flight.[169]

It must be emphasized that the ATLAS IIIA/CENTAUR and ATLAS IIIB/CENTAUR were transitional vehicles. All of them were launched from Pad 36B. Within just a few years they were replaced by Lockheed Martin's new fleet of ATLAS V/CENTAURs. (The latter were launched from Complex 41.) Following the final ATLAS IIIB/CENTAUR launch on 3 February 2005, the deactivation of Complex 36 got underway. Officials had completed a detailed demolition plan for ATLAS facilities in early January 2005, and Launch Complex 36 was secured by the end of July 2005. Lockheed Martin vacated the ATLAS Site Support Building in September and October 2005, and all equipment was removed from that site by the end of November 2005.[170]

AMEC Earth and Environmental, Inc. was the 45th Civil Engineer Squadron's prime contractor for the demolition program, which included TITAN as well as ATLAS/CENTAUR facilities. (The Cape's final TITAN IVB launch occurred at the end of April 2005.) AMEC was awarded two major contracts in 2006 to raze various ATLAS and TITAN sites. On 29 August 2006 AMEC received its Notice to Proceed (NTP) on a $2,569,913 contract to demolish Pad 36B. Officials issued another NTP to AMEC on 22 September 2006, and it allowed the company to begin a $10,215,907 contract to demolish the TITAN Vertical Integration Building, Pad 36A, and ready rooms in power stations 4 and 6. The preconstruction conference for both contracts was held on 9 October 2006. Demolition began in late November 2006, and civil engineering officials expected the work on the ATLAS facilities to continue well into 2007. The final cost of Complex 36's demolition was about $3 million. Not counting their value to the contractor as scrap, the towers on Complex 36 cost $150,000 apiece to pull down and break up. Both Mobile Service Towers (MSTs) were toppled with explosives on Saturday, 16 June 2007, providing a dramatic end to the era of ATLAS, ATLAS II, and ATLAS III operations at the Cape. A total of 145 ATLAS/CENTAUR missions (of all the early types) were launched from Complex 36 between 8 May 1962 and early February 2005.[171]

ATLAS IIIA/CENTAUR and ATLAS IIIB/CENTAUR Commercial Space Operations

ATLAS IIIA/CENTAUR (EUTELSAT W4), AC-201, 24 May 2000

The booster for this mission arrived at Cape Canaveral on 8 March 1999, and it was complemented by Lockheed Martin's ribbon-cutting ceremony for the ATLAS IIIA at Pad 36B on 9 March 1999.[172] At that time, the booster was assigned to the TELSTAR-7 mission, but the

DELTA III/ORION mission failure in May 1999 put an end to that plan. The ATLAS IIIA was grounded pending an investigation of all Pratt & Whitney RL10 engines, and Space Systems Loral pulled the TELSTAR-7 off the ATLAS IIIA and placed the payload on an ARIANE launch vehicle sponsored by the European Space Agency. Following the investigation, Lockheed Martin signed a contract with the European Telecommunications Satellite Organization to launch the EUTELSAT W4 spacecraft on the first ATLAS IIIA. The 7,000-pound satellite was a French-built Alcatel Spacebus 3000, and it was destined to provide communications services for Russia and sub-Saharan Africa.[173]

Technicians erected the ATLAS IIIA/CENTAUR on Pad 36B on 23 February 2000. Officials completed the Simulated Flight Test on 17 March, but the Wet Dress Rehearsal was delayed until 5 April 2000 to allow the contractor time to filter out fuel contaminants. In the meantime, a potential weakness in the CENTAUR's onboard venting system was identified, and testing was required to ensure the hardware met established criteria. Officials rescheduled the spacecraft mate from 5 April to 7 May 2000, and the launch was pushed from 14 April to 15 May 2000. Following flight certification of the onboard venting system on 28 April, preparations for the first launch attempt continued. Officials scrubbed the first launch attempt at 2150Z on 15 May after a radar on Bermuda failed to support. The second launch attempt on 16 May was scrubbed at 2058Z for upper level wind shear. The third launch attempt on 17 May was scrubbed at 2350Z after the CENTAUR's Remote Control Unit (RCU) Master Inhibitor Bus redlined at T minus 30 seconds. The fourth launch attempt on 20 May was scrubbed at 2355Z due to a Computer Controlled Launch Set (CCLS) software issue complicated by the intrusion of two boats in the launch danger area earlier in the countdown. Finally, despite problems with an Uninterruptible Power Supply, Front End Processors, and a fouled range, the fifth launch attempt led to a safe and successful launch of the ATLAS IIIA/CENTAUR at 2310:04.916Z on 24 May 2000. Spacecraft separation occurred as planned, and the mission was successful.[174]

ATLAS IIIB/CENTAUR (ECHOSTAR VII), AC-204, 21 February 2002

The object of the AC-204 mission was to place the ECHOSTAR VII commercial communications satellite in a super-synchronous transfer orbit from Pad 36B. Lockheed Martin Commercial Space Systems (LMCSS) built the 9,200-pound spacecraft for the EchoStar Communications Corporation. As the fourth spacecraft in EchoStar's Dish Network constellation, the three-axis stabilized ECHOSTAR VII was purchased to provide regional Ku-Band video and data relay services to the Continental United States, Alaska, and Hawaii. The payload was processed at Astrotech's space operations facility in Titusville, Florida. Lockheed

Martin Astronautics built the ATLAS IIIB launch vehicle, integrated it with the payload, and launched it. The mission featured the first launch of the ATLAS IIIB/CENTAUR.[175]

Though the launch was tentatively planned for 31 October 2001, the spacecraft was delayed, and its tardy arrival precluded accomplishment of the mission before late February 2002.[176] As the delay continued, spacecraft mating and certification slipped from early December 2001 to early January 2002 to early February 2002. In the meantime, officials completed the Wet Dress Rehearsal on 6 December 2001. Engineers mated the payload to the launch vehicle on 8 February 2002. The Composite Electrical Readiness Test was completed successfully on 11 February 2002. Weather violations kept the range 'Red' from 0735Z to 1053Z during the countdown on the 21 February, and there was one 30-minute-long unscheduled hold at T minus 105 minutes (to correct a liquid nitrogen leak in the CENTAUR). Nevertheless, there were no major instrumentation outages during the countdown, and the ATLAS IIIB/CENTAUR lifted off Pad 36B at 1243:01.102Z on 21 February 2002. The launch was successful.[177]

ATLAS IIIB/CENTAUR (ASIASAT-4), AC-205, 12 April 2003

The object of the AC-205 mission was to place the ASIASAT-4 commercial communications satellite in a super-synchronous transfer orbit from Pad 36B. The spacecraft was launched aboard an ATLAS IIIB/CENTAUR built by Lockheed Martin International Launch Services. The ASIASAT-4 was an HS601HP spacecraft built by Hughes Electronics Corporation and Boeing Satellite Systems Inc. (formerly known as Hughes Space and Communications Company). The spacecraft carried 28 C-Band and 20 Ku-Band transponders, and it was purchased by Asia Satellite Telecommunications Company Limited to provide communications services to users in more than 40 countries.[178] According to the flight scenario, the ATLAS IIIB/CENTAUR chosen for the mission would be launched on a flight azimuth of 96.7 degrees. The spacecraft would be released near the western coast of southern Africa approximately 30 minutes and 53 seconds after lift-off. Subsequently, it would be maneuvered into a geosynchronous orbit of 122 degrees East longitude (i.e., over Borneo). The ASIASAT-4 had an operational life expectancy of 15 years. The combined cost of the ASIASAT-4 and its launch vehicle was $240 million.[179]

The ASIASAT-4 arrived at Cape Canaveral on 14 February 2003, and the launch was scheduled for 11 April 2003. The Wet Dress Rehearsal was completed on 19 March, and engineers mated the spacecraft to its launch vehicle on 1 April 2003. Officials conducted the Combined Systems Test on 2 April, and the Launch Readiness Review was completed on 8 April 2003. An intermittent Command Message Encoder Verifier (CMEV) fault proved to be a nuisance for officials, but "gusty surface winds" were a more serious cause for concern during the first launch attempt on 11 April 2003. Indeed, the countdown on 10/11 April experienced a 50-minute-long unscheduled hold (at T minus five minutes), and the launch was scrubbed at 0054Z on the 11th for upper level winds. The countdown on 11/12 April 2003 went well, though a Cinetheodolite at a camera site was slightly out of focus due to a faulty tensioner spring. Other items were minor, and the ATLAS IIIB/CENTAUR lifted off Pad 36B at 0047:01.143Z on 12 April 2003. The launch was successful.[180]

ATLAS IIIA/CENTAUR (MBSat), AC-202, 13 March 2004

The object of the AC-202 mission was to place the MBSat commercial communications satellite in an elliptical transfer orbit from Pad 36B. The spacecraft was launched aboard an ATLAS IIIA/CENTAUR built by Lockheed Martin International Launch Services. Space Systems Loral built the 9,100-pound MBSat based on Loral's "space-proven" three-axis, body-stabilized 1300 bus. In a joint effort, the Mobile Broadcast Corporation of Japan and SK Telecom of South Korea purchased the MBSat to provide high-quality audio, video, and data services throughout Japan. The combined cost of the spacecraft and launch vehicle was approximately $150 million. According to the flight scenario, the ATLAS IIIA/CENTAUR chosen for the mission would be launched from Pad 36B on a flight azimuth of 97.5 degrees. The spacecraft would be released near the western coast of southern Africa approximately 30 minutes and 29 seconds after lift-off. Subsequently, it would be maneuvered into geosynchronous orbit at 144 degrees East longitude (i.e., over the equator south of Japan and South Korea).[181]

Technicians erected the ATLAS IIIA booster on 15 January 2004. Officials completed the Simulation Flight on 18 February 2004. An electrical problem prevented completion of the Wet Dress Rehearsal (WDR) on 24 February, and bad weather delayed the WDR an additional day. Officials completed the WDR on 26 February, and engineers mated the MBSat to the launch vehicle on 4 March 2004. Though the launch was scheduled for 12 March, an administrative issue involving the spacecraft compelled officials to reschedule the event for 13 March about three days before the actual launch. There were no unplanned holds during the countdown, and the ATLAS IIIA/CENTAUR lifted off Pad 36B at 0540:00.844Z on 13 March 2004. The launch was successful.[182]

ATLAS IIIB/CENTAUR Military Space Operations

ATLAS IIIB/CENTAUR (UHF F-11), AC-203, 18 December 2003

The object of the AC-203 mission was to place the U.S. Navy's UHF F-11 communications spacecraft in a super-synchronous transfer orbit from Pad 36B. The spacecraft was launched aboard an ATLAS IIIB/CENTAUR built by Lockheed Martin International Launch Services. Boeing Satellite Systems Inc. built the UHF F-11 spacecraft. The 6,545-pound spacecraft was equipped to handle 41 Ultra High Frequency (UHF) channels and 20 Extremely High Frequency (EHF) channels. Like other satellites in the Navy's UHF Follow-On satellite constellation, the UHF F-11 supported U.S. military communications worldwide.[183] According to the flight scenario, the ATLAS IIIB/CENTAUR chosen for the mission would be launched on a flight azimuth of 96.63 degrees. The spacecraft would be released near the western coast of southern Africa approximately 30 minutes and 53 seconds after lift-off. Subsequently, it would be maneuvered into a geosynchronous orbit.[184]

The launch was tentatively scheduled for 12 November 2003. Unfortunately, "spacecraft issues" arose in the summer of 2003, and officials decided to move the launch to 2 December 2003. A payload processing schedule delay (noted in mid-October 2003) prompted more rescheduling, and officials ultimately planned to launch the UHF F-11 on 16 December 2003. As preparations for the launch continued, officials conducted the Wet Dress Rehearsal on 23 October 2003. Engineers mated the UHF F-11 spacecraft to the vehicle on 5 December, and the Combined Electrical Readiness Test was completed successfully on 8 December 2003. Instrumentation "show-stoppers" were rare occurrences in the Eastern Range's launch history, so one of the downrange radars caused some anxiety when it went Non-Mission Capable (NMC) on 10 December 2003. The radar was unable to operate due to a faulty RF Amp Pulse Transformer in its transmitter. Without the radar, the range could not support the UHF F-11 launch. There was some trouble transporting the replacement part (e.g., a Klystron tube) downrange on 12 December, but the contractor completed the replacement operation successfully over the weekend of 13/14 December 2003. The radar was functioning properly on 15 December 2003. The Eastern Range was ready, and the first launch attempt got underway on 15/16 December 2003 as planned.[185]

Unfortunately, an equipment anomaly surfaced during the countdown on the 16th, and it prompted officials to scrub the launch on that date. The ATLAS liquid oxygen Fill and Drain "bolt cutter" registered excessively low temperature (lower than the –180 degree limit). Authorities couldn't launch the vehicle, so the launch was scrubbed and recycled for 18 December 2003.[186]

The second countdown on 17/18 December went well, and the ATLAS IIIB/CENTAUR lifted off Pad 36B at 0224:00.000Z on 18 December 2003. The launch was successful, but there were some significant instrumentation anomalies involving the new Flight Operations Version 1 range safety system. The FOV-1 system functioned well enough to assume the "primary" role for

Mission Flight Control Officer (MFCO) display, but "legacy" range safety/range control systems were the primary and secondary systems for High Density Data (HDD) output. The Distributed Range Safety Display (DRSD) and FOV-1A served as backup systems for HDD output. Put simply, "glitches" in the new systems would have to be worked out before Range Safety could relegate the legacy systems to a backup role.[187]

ATLAS IIIB/CENTAUR (NROL-23), AC-206, 3 February 2005

The object of the AC-206 mission was to place a National Reconnaissance Office satellite (NROL-23) in a transfer orbit from Pad 36B. The flight marked the sixth and final launch of the ATLAS III and the 145th and final ATLAS launch from Complex 36. An ATLAS IIIB/CENTAUR was chosen for this classified mission. According to the flight scenario, the vehicle was launched on a flight azimuth of 43.02 degrees. The payload separated from the launch vehicle approximately 79 minutes after lift-off.[188]

Technicians erected the ATLAS IIIB booster on Pad 36B on 1 December 2004. Engineers mated the CENTAUR upper stage on 3 December, and officials completed the launch vehicle's Initial Current Profile and Compatibility Test (ICPACT) on 20 December 2004. During a preliminary run of the ICPACT on 10 December, analysts noted an apparent internal short in the CENTAUR's remote control unit. Further sleuthing revealed the short was caused by excessive solder debris, and the condition was blamed on the supplier's "poor workmanship." A new flight circuit arrived on 24 December, and engineers installed it on 3 January 2005 to correct the remote control problem. A portion of the ICPACT verified a good circuit on 3 January, and officials ran a simulation flight (SIMFLIGHT) successfully between 7 and 11 January 2005. Officials completed the Wet Dress Rehearsal (WDR) on 14 January 2005. Engineers mated the payload to the launch vehicle on 17 January, and officials completed the Combined Electrical Readiness Test (CERT) on 19 January 2005. Preparations continued for the launch, which was scheduled for 27 January 2005.[189]

Inspectors noticed some RP-1 (fuel) around the base of the first stage on 20 January 2005, and they traced this "seepage" (e.g., nine cubic centimeters) to the lower mating ring on the first stage's RP-1 tank. Pressure testing, de-tanking, and x-ray procedures were completed, and the launch was put on hold until the matter could be resolved. The source of the seepage was located precisely, and the entire seam was thoroughly cleaned and resealed with polyurethane sealant (except for two small areas where tubes had to be installed to drain off any subsequent seepage). As an additional precaution "absorbent pads" were installed over the seam where seepage had been detected, and the mission was flown with the pads in place. Authorities cleared the vehicle for flight on 27 January, and range officials rescheduled the launch for 3 February 2005.[190]

There were no unplanned holds during the countdown, and the ATLAS IIIB/CENTAUR lifted off Pad 36B at 0741:00.931Z on 3 February 2005. The launch was successful, and it marked the end of a string of ATLAS launches on Complex 36 extending all the way back to the very first ATLAS/CENTAUR launch at Cape Canaveral on 8 May 1962.[191]

ATLAS V/CENTAUR Overview

The Lockheed Martin Evolved Expendable Launch Vehicle (EELV) Program

After many years of government-funded studies, the Air Force developed a "roadmap" in 1994 for the acquisition of Evolved Expendable Launch Vehicles (EELVs) to succeed the current generation of DELTA IIs, ATLAS IIs, and TITAN IVs at Cape Canaveral.[192] The Space and Missile Systems Center (SMC) released a Request for Proposal in May 1995, and four companies (e.g., Alliant Techsystems, Boeing, Lockheed Martin, and McDonnell Douglas) were each awarded 15-month-long contracts in August 1995 to elaborate their concepts for the EELV.[193] On 20 December 1996, two of the aforementioned contractors — Lockheed Martin and McDonnell Douglas — were each awarded a $60 million contract to complete independent 17-month-long Pre-Engineering & Manufacturing Development studies for the EELV. Boeing bought McDonnell Douglas Aerospace in August 1997. Consequently, it was Boeing that proposed a new family of "spacelifters" based on McDonnell Douglas' DELTA IV launch vehicle design. Lockheed Martin also proposed its version of the EELV, which became known as the ATLAS V.[194]

The U.S. Government planned to award two separate $800,000,000 contracts for the development, manufacture, and deployment of *a single family* of medium and heavy EELVs in June 1998. However, Pentagon officials eventually approved the new plan to encourage the development of *two families* of EELVs for government and commercial space missions in the near future. On 16 October 1998, Acting Secretary of the Air Force F. Whitten Peters announced the award of two $500,000,000 contracts to Boeing and Lockheed Martin for Engineering and Manufacturing Development (EMD) of their respective EELV families. At the same time, Boeing was awarded another contract worth $1.38 billion for 19 government-sponsored launches between Fiscal 2002 and the end of Fiscal 2006. For its part, Lockheed Martin was awarded a $650,000,000 contract for nine government-sponsored launches between Fiscal 2002 and the end of Fiscal 2005. Spurred on by the commercial market as well as government interest, Boeing and

Lockheed Martin became quite keen to develop EELVs to meet a wide range of missions at lower cost.[195]

Lockheed Martin initially proposed a family of EELVs consisting of small, medium, and heavy vehicles based on a Common Core (CC). The CC would be a self-supporting (non-pressurized) aluminum first stage, slightly wider than the first stage on earlier ATLAS vehicles. It would be powered by an RD-180 liquid oxygen/kerosene engine, which was based on the Russian RD-170 first stage engine used on the ZENIT space launch vehicle.[196] The small variant was soon dropped, and Lockheed Martin focused its development efforts on a middle-sized EELV with considerable growth potential. It was called the HLV-A, and it employed a CENTAUR upper stage powered by an RL-10A-4 engine rated at 22,000 pounds of thrust.[197]

Lockheed Martin proposed to use Launch Complex 41 at Cape Canaveral for its EELV launch operations. Toward that end, Lockheed Martin hired AJT & Associates (Cape Canaveral) to draw up new plans for the launch pad and contracted Hensel Phelps Construction Company (Greeley, Colorado) to modify the pad once the government granted Right of Entry (ROE). In Lockheed's vision of its future EELV operations, payloads would be processed and encapsulated off-site before being mated to the new vehicles in a vertical processing building. As part of Lockheed Martin's "clean pad" launch philosophy, the completed vehicle and its payload would then be transported to the launch pad where it could be fueled and launched in less than 24 hours.[198]

Lockheed Martin's proposal came at a very good time. Several years earlier, the Air Force decided the TITAN IV launch rate at the Cape did not justify operating Complex 41 after Fiscal 1999. Initially, the plan was to deactivate the pad and "mothball" it as a hedge against a catastrophic launch failure on Complex 40. By the end of 1995, other options including complete demolition of the site were considered, but Detachment 8 of the Space and Missile Systems Center and the 45th Space Wing agreed that nothing substantive could be accomplished until EELV plans were "firmed up." The EELV selection process identified Lockheed Martin as one of two EELV finalists on 20 December 1996, and planning began again. By August 1997, the Air Force's position was that Complex 41 would be deactivated "immediately following the final TITAN IV launch from the pad" (i.e., sometime in 1999). The pad *would not be placed in caretaker status* even if an EELV contractor decided not to use the site. Aerospace Ground Equipment (AGE) and Real Property Installed Equipment (RPIE) would be identified and removed by Detachment 8 and the 45th Civil Engineer Squadron respectively.[199] A notable exception to the plan was the disposal of Mobile Service Tower (MST). In September 1997, Cape Engineering decided to fold the MST demolition into the EELV contract and have the

EELV contractor sell the scrap to help pay the cost of demolition. The 45th Maintenance Squadron was directed to manage the removal of the pad's communications equipment, and the Eastern Range photographic services contractor would remove the site's camera equipment under the supervision of the 45th Communications Squadron. If Complex 41 became an EELV launch pad, about $5,000,000 would be needed to clean up PCB contamination in areas disturbed by new construction. The clean-up operation would be financed by AFSPC and supervised by the 45th Civil Engineer Squadron. Responsibilities were firmed up in a Memorandum of Agreement dated 6 August 1998.[200]

Since Complex 41 extended into the Merritt Island Launch Area (MILA) relatively close to Shuttle complexes 39A and 39B, the General Counsel to the Secretary of the Air Force (SAF/GCN) decided to draft a permit for NASA's approval. The permit would clarify Air Force/EELV contractor intentions and confirm NASA's support of the Air Force's decision to lease the complex to Lockheed Martin for its EELV program.[201] Comments on the drafted permit were solicited in August 1998, and William W. Brubaker (NASA's Director of Facilities Engineering) signed the permit on 23 October 1998. The permit would be in force from 1 November 1998 through 31 December 2021. Mr. Brubaker indicated in his cover letter to the signed agreement that "all NASA approval requirements for leasing and construction have been removed." The Air Force only had to 1) provide NASA an opportunity to review EELV plans, 2) give "full consideration" to NASA's comments, and 3) coordinate construction details with the Kennedy Space Center's Master Planning and Siting Office so NASA would have a record of changes in the site's description. With NASA's approval, the Air Force was free to turn Complex 41 over to Lockheed Martin for EELV operations.[202]

Other excess facilities were allocated to Lockheed Martin and Boeing for their EELV programs in 1998. In late January 1998, the Space and Missile Systems Center (SMC/MVK) approved Lockheed Martin's request to use the Motor Inert Storage (MIS) building for its EELV operations. In December 1997, the Receipt Inspection Storage (RIS) Building was allocated to Lockheed Martin for EELV general storage purposes, but only until someone with Solid Rocket Motor (SRM) processing requirements came along to put the facility to better use. In early February 1998, Boeing asked to use the RIS and Segment Ready Storage (SRS) buildings for its EELV program, and the company provided more information in June on its intended use of the Receipt Inspection Storage (RIS) Building. The facility would be used for the processing and staging of DELTA IV Heavy separation motors as well as GEM-60 Solid Rocket Motors. In early July 1998, the 45th Space Wing Commander asked the Space and Missile Systems Center to remove the RIS and SRS buildings from Lockheed Martin's list of facilities so they could be licensed to Boeing for its EELV program. Boeing agreed to assume all operations, maintenance, and modification costs for both facilities. The Space and Missile Systems Center and Air Force Space Command both agreed, and the reallocation was approved in mid-July 1998.[203]

On 8 September 1998, Brigadier General F. Randall Starbuck approved a comprehensive list of Cape facilities to be allocated to Boeing and Lockheed Martin for their respective EELV programs. For Boeing, the list included Complex 37, a site for the company's new Horizontal Integration Facility (HIF), the CENTAUR Processing Facility (CPF), facilities in the Hangar C area, the RIS Building, and the SRS Building. Lockheed Martin's share of the list included Complex 41, Hangar J, the Missile Inert Storage building, rooms in the Launch Operations Control Center, and various 29100-series facilities. Boeing received a Right of Entry (ROE) to Complex 37 on 8 June 1998. Following receipt of its Right of Entry on 24 October 1998, Lockheed Martin moved its construction trailers onto Complex 41 and started clearing and grubbing for the new Vertical Integration Facility (VIF). Hensel Phelps served as the EELV subcontractor for Lockheed Martin, and the company also handled TITAN IV deactivation work under Titan-0012 Contract Mod KX-0143. Work was solidly underway at both EELV launch sites at the close of 1998.[204]

The imminent prospect of EELV operations at the Cape prompted the government to spell out how it would supervise the new programs. In this, the Air Force departed from the traditional oversight policy it had enforced for many decades. Under the EELV Concept of Operations (CONOPS) signed in early September 1997, the EELV contractor was given total system responsibility and retained ownership of all EELV flight hardware. The contractor would also maintain all critical ground support equipment and facilities. Air Force Space Command would determine the most cost-effective way to maintain *non-critical* ground support equipment, and who — contractor, Air Force or third party — would maintain it. The EELV contractor would be responsible for manufacturing, processing, and launching the vehicle. As a result, the U.S. Government's role would shift from oversight to *insight*. This shift was designed to reduce bureaucratic layering and get the Air Force out of the "product testing" business. Nevertheless, Air Force Materiel Command would retain sufficient insight to ensure EELVs were flight-worthy, and Air Force Space Command would ensure the safety of the EELV in ground and flight operations after the vehicle arrived at the Cape. The 45th Space Wing Commander remained the Launch Decision Authority (LDA) and made the final decision to launch. Though the EELV contractor was expected to have his own system safety program, the Wing Commander had to ensure that operational hazards were identified and safety risks were minimized. As was true of earlier launch programs, the EELV contractor had to obtain Air Force Space Command's approval for all EELV flight plans and launch operations.[205]

Deactivation of Complex 41 and Construction for ATLAS V

The ATLAS V program achieved a major milestone at Cape Canaveral on 24 October 1998 when the U.S. Government granted Lockheed Martin its Right of Entry (ROE) to Complex 41. In support of the overall effort, the 3rd Space Launch Squadron and 45th Civil Engineer Squadron worked with ten separate contractors and representatives from AFSPC, AFMC and the NRO to complete Complex 41's deactivation. As part of the deactivation process, technicians flushed all fuel lines on the complex, and workers shipped out approximately 20 train carloads of soil for sanitizing and reuse. On 14 October 1999, the Olshan Demolishing Company used 180 pounds of explosives to topple the site's old TITAN IV Mobile Service Tower (MST) and Umbilical Tower (UT). Workers gleaned about $2.5 million worth of salvage from the complex before the event, and they removed approximately 8 million pounds of steel following the demolition. In all, 72 days were required to completely decommission the site's hypergolic propellant systems.[206] Officials deactivated the complex 42 days ahead of schedule.[207]

With a workforce of nearly 500, Lockheed Martin and its assistant contractor, Hensel Phelps, completed facility modifications for the ATLAS V in the spring of 2001. On the Vehicle Integration Facility (VIF) site 1,800 feet south of Complex 41, workers poured more than 1,500 cubic yards of concrete on 27 March 1999 to create the VIF slab.[208] By the end of 1999, construction reached the 250-foot level of the 292-foot VIF structure. Construction was also underway on the Entry Control Building (ECB) and the HVAC (Air-conditioning and Humidity Control) Shelter, just south of the VIF. Workers "topped off" the VIF in early March 2000, and they started building the new Mobile Launch Platform (MLP) south of the VIF during the summer of 2000.[209] Work was underway in 2000 to renovate the Missile Inert Storage (MIS) Building to accommodate the ATLAS V Launch Control Center (LCC) and Mission Director's Center (MDC). Upon completion, company officials renamed the MIS the ATLAS V Spaceflight Operations Center (ASOC).[210] Workers delivered and installed a 42,000-gallon liquid hydrogen tank and two 45,000-gallon stainless steel RP-1 fuel tanks in the fall of 2000. A 465,000-gallon liquid oxygen spherical tank was in place before the end of the year. By the time the first ATLAS V arrived in early June 2001, Complex 41 was ready to begin pathfinder operations.[211]

Unlike Boeing, Lockheed Martin decided to use its first Evolved Expendable Launch Vehicle (designated AV-001) as both a pathfinder and a launch vehicle. The first ATLAS V booster and upper stage arrived at the Cape during the first week of June 2001. After several weeks of booster and facility tests in the ASOC, technicians moved the ATLAS V to the VIF in October 2001. They stacked the vehicle on the Mobile Launch Platform, and they added a payload simulator to the vehicle in early November 2001. Following additional facility tests in the VIF, workers de-stacked the ATLAS V and returned it to the ASOC for final horizontal checkout and avionics installation in December 2001. Technicians erected AV-001 again on 22 February 2002 and the *operational* AV-001 was rolled out to Complex 41 on 6 and 7 March 2002.[212]

The ATLAS V/CENTAUR Family of Launch Vehicles

The ATLAS V/CENTAUR (400 and 500 variants) constituted Lockheed Martin's current family of modern ATLAS space launch vehicles. The ATLAS V (400) made its debut launch from Complex 41 on 21 August 2002. The "400" used a single-stage, kerosene-fueled ATLAS RD-180 engine to power a 12.5 x 106.5-foot Common Core Booster (CCB). Though the RD-180 provided 860,200 pounds of thrust all by itself, the basic "400" could be equipped with up to three strap-on Aerojet Solid Rocket Motors (SRMs) to lift a variety of heavier payloads.[213] In either configuration, the "400" employed payload fairings that had been developed for the ATLAS II and ATLAS III families. The ATLAS V (500) extended the capabilities of the ATLAS V by incorporating up to five SRMs and a larger 5.4-meter diameter payload fairing.[214]

For its heaviest payloads (e.g., up to 28,660 pounds to Geosynchronous Transfer Orbit), Lockheed Martin intended to combine three CCBs to form the ATLAS V-Heavy. For their upper stage, all ATLAS Vs used a 10 x 41.6-foot Pratt & Whitney CENTAUR based on the CENTAUR RL10A-4-1 used on the ATLAS IIIB/CENTAUR. The ATLAS V's CENTAUR upper stage (equipped with either one or two Pratt & Whitney RL10A-4-2 engines) was designated the "Common Centaur." In its tallest configuration, the ATLAS V (400) stood about 194 feet tall. An ATLAS V (500) equipped with a 77-foot-long fairing was 205 feet tall. If an ATLAS V-Heavy were ever built and launched, it would stand 215 feet tall with its 87-foot-long payload fairing in place.[215]

In a typical ATLAS V/CENTAUR flight scenario (400 variant), the launch vehicle's RD-180 ignited about three seconds before lift-off. Rising vertically for about 250 feet, the vehicle rolled into its flight azimuth and began its initial roll, pitch, and yaw program under the control of the CENTAUR's inertial guidance system. Once its fuel was depleted (e.g., about four to four and one-half minutes into the flight), the RD-180 shut down, and the booster separated from the CENTAUR upper stage and payload. The CENTAUR fired about ten seconds after separation from the booster. The payload fairing was jettisoned during the ATLAS boost phase or during the CENTAUR's first "burn." In either event, the first CENTAUR burn (lasting from about five minutes to nearly 14 minutes depending on the configuration of vehicle) injected the upper stage and spacecraft into a "moderately eccentric" parking orbit. Following a coasting period lasting from eight minutes to more than an hour (e.g., the RAINBOW-1, AMC-16 and ASTRA 1 KR missions), the CENTAUR restarted and fired to place the payload in a supersynchronous transfer orbit. The spacecraft separated from the upper stage some minutes later, and it used its onboard propulsion system to achieve final orbit.[216]

Range instrumentation for a typical ATLAS V/CENTAUR mission included radars at the Cape, Patrick AFB, Jonathan Dickinson Missile Tracking Annex, and the island of Antigua. Command transmitter/exciters at the Cape, Jonathan Dickinson, and Antigua provided command and control. HH-60 helicopters operated out of Patrick AFB to provide safety sea surveillance, and an SCE Learjet provided weather reconnaissance. Telemetry units on Merritt Island, Jonathan Dickinson, and Antigua supported the flights, and so did the Centralized Telemetry Processing System (CTPS) at Cape Canaveral. Optical support included the Distant Object Attitude Measurement System (DOAMS), the Advanced Transportable Optical Tracking System (ATOTS) at the Cape, and cameras at other range sites to ensure at least four optical systems covered each launch. Patrick's WSR-74C weather radar, Melbourne's WSR-88D radar, and a whole host of weather instrumentation systems at Cape Canaveral supported each flight.[217]

ATLAS V/CENTAUR Commercial Space Operations

ATLAS V/CENTAUR (EUTELSAT HOT BIRD 6), AV-001, 21 August 2002

The object of the AV-001 mission was to place the EUTELSAT HOT BIRD 6 commercial communications satellite in a super-synchronous transfer orbit from Pad 41. Significantly, the mission was also the inaugural flight of the ATLAS V (400) Evolved Expendable Launch Vehicle (EELV). Alcatel built the 8,600-pound HOT BIRD 6 spacecraft for the Eutelsat Corporation. The satellite was based on Alcatel's Spacebus 3000B3 design, and it was equipped with 28 Ku-Band and four Ka-Band transponders to provide digital television and radio broadcast services to Eutelsat's customers in Europe, North Africa, and parts of the Middle East. The payload was the "most powerful satellite" launched for Eutelsat, and it had an operational life expectancy of more than 12 years. The HOT BIRD 6 was processed at Astrotech's space operations facility in Titusville, Florida. Lockheed Martin Astronautics built the ATLAS V/CENTAUR launch vehicle, integrated it with the payload, and launched it.[218]

Lockheed Martin decided to use the AV-001 vehicle as both a "pathfinder" and a launch vehicle. The first ATLAS V booster arrived at the Cape during the first week of June 2001. After spending several weeks at the ATLAS V Spaceflight Operations Center (ASOC), the ATLAS V was moved to the Vehicle Integration Building in October 2001. Contractor personnel stacked the vehicle on the Mobile Launch Platform, and they added a payload simulator to the ATLAS V in early November 2001. Engineers completed their tests and de-stacked the vehicle by 14 November, and they returned AV-001 to the ASOC for additional checks. As preparations for the first ATLAS V/CENTAUR mission got underway, the successful flight of the first ATLAS IIIB/CENTAUR on 21 February 2002 was encouraging — it featured the successful debut of the "stretched" CENTAUR on which the ATLAS V's upper stage was based. Technicians erected AV-001 (again) on 22 February 2002, and the *operational* AV-001 was rolled out to Complex 41 on 6 and 7 March 2002. Officials conducted a Wet Dress Rehearsal (WDR) in mid-March, and the launch campaign WDR was completed 15-17 July 2002. The launch was scheduled for 12 August, but it slipped to indefinite status on 26 July to give the contractor additional time to perform some final tests. (Officials settled on a firm date of 21 August 2002 a few days later.) Engineers mated the HOT BIRD 6 to the vehicle on 9 August 2002 following spacecraft and payload fairing operations at Astrotech's new Lee processing facility in Titusville.[219] Officials completed the Integrated System Test (IST) on 13 August 2002. Throughout the entire campaign, the 45th Space Wing's safety organization provided oversight for launch and payload processing, facility and ground support equipment designs, and over 200 first-time hazardous and safety-critical processing procedures.[220]

In keeping with Lockheed Martin's streamlined new launch system, the ATLAS V/CENTAUR was delivered to the launch pad only one day before the launch. (Typically, ATLAS IIA/CENTAURs needed more than a month at the launch pad before a countdown, but most of the final preparations for the ATLAS V could be made *before* the completed vehicle and its payload went to the pad.) The local weather cooperated, and the countdown proceeded smoothly. There were no unscheduled holds, and the ATLAS V/CENTAUR lifted off Complex 41 at 2205:00.193Z on 21 August 2002. The $120 million HOT BIRD 6 spacecraft entered the proper orbit approximately 31 minutes later. In what Lockheed Martin characterized as a "flawless flight," all systems performed properly. As the first operational flight of the ATLAS V, the success of the HOT BIRD 6 mission validated the launch vehicle and all the new ground facilities Lockheed Martin built for its new 'clean pad' concept of operations. The AV-001 launch was a major milestone for the Cape as well.[221]

ATLAS V/CENTAUR (HELLAS-SAT), AV-002, 13 May 2003

The object of the AV-002 mission was to place the HELLAS-SAT commercial communications satellite in a super-synchronous transfer orbit from Complex 41. Alcatel built the 7,603-pound

HELLAS-SAT spacecraft for HELLAS-SAT Consortium, Ltd. (Nicosia, Cyprus). The satellite was an Astrium Eurostar 2000-Plus model, and it was equipped with 30 Ku-Band transponders to provide voice, internet, and broadcast digital television services to HELLAS-SAT's customers in Europe. According to International Launch Services (the launch provider), the HELLAS-SAT was "the first telecommunications satellite for Greece and Cyprus." The spacecraft had an operational life expectancy of 15 years. The HELLAS-SAT was processed at Astrotech's space operations facility in Titusville, Florida. Lockheed Martin Astronautics built the ATLAS V/CENTAUR (400) launch vehicle chosen for the mission. Lockheed also integrated the ATLAS V with the payload and launched it.[222]

The HELLAS-SAT arrived at Cape Canaveral on 11 February 2003, and technicians transported the spacecraft to Astrotech's facility in Titusville, Florida shortly thereafter. Engineers encapsulated the satellite in its payload fairing on 30 March 2003. As payload processing continued, technicians transported the ATLAS V/CENTAUR launch vehicle from the Vertical Integration Facility (VIF) to the launch pad on Complex 41. Officials completed two Wet Dress Rehearsals (WDRs) on 19 February and 29 April 2003 respectively. Engineers mated the HELLAS-SAT to the launch vehicle on 2 May 2003. Officials completed the Combined Electrical Readiness Test and Launch Readiness Review successfully on 10 May 2003. Though the launch was scheduled for 12 May 2003, it was pushed to 13 May due to a voltage problem in the CENTAUR's remote data unit. Though two potentially major instrumentation problems arose during the countdown on the 13th, they both proved to be minor in nature.[223] As the countdown continued, there was one unscheduled hold lasting 13 minutes, and it involved two separate issues: 1) a range boat in the "one boat contour" (i.e., a fouled range incident) and 2) a faulty hydrogen sensor reading on the launch vehicle. Both conditions cleared, and the ATLAS V/CENTAUR lifted off Complex 41 at 2210:00.855Z on 13 May 2003. According to International Launch Services (ILS), the ATLAS V injected the HELLAS-SAT into orbit very close to pre-planned parameters. The mission was successful.[224]

ATLAS V/CENTAUR (RAINBOW-1), AV-003, 17 July 2003

The object of the AV-003 mission was to place the 9,445-pound RAINBOW-1 commercial communications spacecraft in a geosynchronous transfer orbit from Complex 41. Lockheed Martin Space Systems Company built the RAINBOW-1 (Model A2100) spacecraft for Cablevision, a direct-to-home broadcast service company. The satellite carried 36 Ku-Band transponders and 26 individually programmable "spot beams." The mission's ultimate goal was to provide Cablevision services to customers in the Continental United States and Canada. The launch contractor, International Launch Services (ILS),[225] chose an ATLAS V (521) vehicle for the mission.[226] The larger fairing was required to shield the RAINBOW-1's "sophisticated" antenna array. According to the flight scenario, the vehicle would be launched on a flight azimuth of 86.1 degrees. During the flight, the CENTAUR upper stage was programmed to accomplish two burns with a 77-minute-long 'coasting' period between them. If all went well, the RAINBOW-1 spacecraft would separate from the vehicle about one hour, 40 minutes, and 20 seconds after lift-off. The satellite's final geosynchronous orbital location was 61.5 degrees West longitude.[227]

Engineers accomplished initial power-on testing for the RAINBOW-1 launch vehicle at the ATLAS Spaceflight Operations Center (ASOC) in late April/early May 2003. Technicians attached two Aerojet solid rocket motors to the vehicle on 23 and 24 May, and officials completed a readiness test on 10 June 2003. The Wet Dress Rehearsal was accomplished on 24 June, and the spacecraft was mated to the launch vehicle on 7 July 2003. The launch was scheduled for 17 July 2003, and officials were worried about "afternoon thunderstorms" before the launch. As the countdown progressed, the T minus 4-minute built-in hold had to be extended for weather as well as a booster helium leak. That delay aside, the countdown continued. The ATLAS V/CENTAUR lifted off Pad 41 at 2345:01.000Z on 17 July 2003. The launch was completely successful. According to International Launch Services, the CENTAUR upper stage released the RAINBOW-1 into a "nearly perfect transfer orbit."[228]

ATLAS V/CENTAUR (AMC-16), AV-005, 17 December 2004

The object of the AV-005 mission was to place the AMC-16 commercial communications satellite in a transfer orbit from Complex 41. Lockheed Martin built the AMC-16 based on the A2100 spacecraft platform. The spacecraft weighed 4,065 kilograms, and it had an operational life expectancy of 15 years. It featured twenty-four 36-MHz Ku-Band transponders and twelve 125-MHz Ka-Band spot beams. The AMC-16 was owned by SES AMERICOM and leased to EchoStar Communications Corporation to enlarge the corporation's fleet of commercial satellites and provide additional high-definition television channels to local U.S. markets. Lockheed Martin Astronautics employed an ATLAS V 521 launch vehicle featuring two Aerojet Solid Rocket Motors and a five-meter payload fairing to accomplish the mission.[229]

The ATLAS V booster selected for the AMC-16 mission arrived at Cape Canaveral on 25 September 2004, just 24 hours before Hurricane JEANNE made landfall. Though JEANNE did not damage the launch vehicle, erection of the booster was delayed due to the hurricane's impact. Consequently, Lockheed Martin requested a change in the launch date from 6 December to 16 December 2004, and officials approved that change on 5 October 2004. The booster was erected on 19 October, and engineers installed the CENTAUR upper stage on 26 October 2004. Following a slight delay due to a trailer misalignment, engineers finished attaching both Aerojet Solid Rocket Motors to the vehicle on 25 October 2004.[230]

Officials completed the Launch Vehicle Readiness Test on 15 November, and they completed the Wet Dress Rehearsal (WDR) on 23 November 2004. On 3 December officials slipped the launch date to 17 December 2004. Engineers mated the AMC-16 to the launch vehicle on 6 December, and officials completed an Integrated System Test on 8 December 2004. Preparations continued for the launch on the 17th.[231]

There were two unplanned holds during the countdown on 17 December. The T minus 120 minutes hold was extended to 90 minutes to allow the user time to complete ground support actions. The second unplanned hold occurred at T minus 2 minutes, and it continued for 56 minutes while a redline monitor fault cleared and winds aloft improved. Lockheed Martin entered the terminal countdown with an FAA waiver (with Air Force concurrence) for a CBC Flight Termination System (FTS) battery temperature out of tolerance, but the vehicle lifted off without incident. Lift-off occurred at 1206:59.352Z on 17 December 2004. The launch was successful.[232]

ATLAS V/CENTAUR (INMARSAT-4), AV-004, 11 March 2005

The object of the AV-004 mission was to place the first INMARSAT-4 commercial communications satellite in a transfer orbit from Complex 41.[233] The satellite was the first of a new generation of communications satellites built by EADS Astrium (a European company) for Inmarsat, a London-based corporation.[234] The 4-series spacecraft was 60 times more powerful that its 3-series predecessors, and it was designed to deliver 3G-compatible broadband data service to mobile users. The 13,138-pound spacecraft was launched on an ATLAS V/CENTAUR vehicle equipped with a 4-meter diameter payload fairing, three Aerojet Solid Rocket Motors (SRMs), and a single-engine CENTAUR upper stage. The INMARSAT-4 was touted as the "heftiest commercial communications payload launched from Cape Canaveral," and the ATLAS V selected for the mission was equipped with the largest number of SRMs employed up to that time.[235]

According to the flight scenario, the ATLAS V/CENTAUR was launched on a flight azimuth of 90 degrees, and it released the spacecraft over Africa approximately 32 minutes and 16 seconds after lift-off. The extremely high transfer orbit (e.g., 56,270 x 274 miles) took advantage of the launch vehicle's great power, and it saved fuel as ground controllers maneuvered the

INMARSAT-4 into its final geostationary orbit over the equator at an altitude of 22,300 miles. Positioned at 64 degrees (over the Indian Ocean), the INMARSAT-4 would provide services to customers in Europe, Africa, the Middle East, and Asia.[236]

Technicians erected the booster on Complex 41 on 11 January, and engineers installed the upper stage on 19 January 2005. Initial attempts to install the SRMs were delayed by high winds and minor hardware problems, but technicians completed all SRM attachments by 10 February 2005. Officials accomplished the Launch Vehicle Readiness Test (LVRT) on 10 February, and the Wet Dress Rehearsal (WDR) was completed successfully on 22 February 2005. Engineers mated the INMARSAT-4 payload to the launch vehicle on 2 March, and preparations continued for a launch on 10 March 2005. The first countdown was scrubbed at 2154Z on 10 March for two problems: 1) the Flight Termination System experienced an intermittent pilot tone drop-out, and 2) the upper booster liquid oxygen line Pogo Accumulator indicated wetting as the liquid oxygen tank was pressurized. Both items were resolved quickly, and the launch was rescheduled for 11 March 2005.[237]

The countdown on the 11th went well. There were no unplanned holds during the countdown, and the ATLAS V/CENTAUR lifted off Complex 41 at 2142:00.280Z on 11 March 2005. The flight was nominal, and the payload was placed in the proper transfer orbit. The $250 million mission was the latest in a string of 76 consecutive ATLAS launch successes dating back to 1993. Lockheed Martin looked forward to even greater accomplishments in the future.[238]

ATLAS V/CENTAUR, ASTRA 1KR, 20 April 2006

The object of this ATLAS V/CENTAUR mission was to boost the ASTRA 1KR commercial television broadcasting spacecraft into orbit from Complex 41. The satellite was purchased to join Luxembourg's SES ASTRA constellation of 13 direct-to-home broadcast spacecraft. Weighing in at 4,200 kilograms, the ASTRA 1KR was based on Lockheed Martin's popular A2100AX communications platform. The satellite carried 32 active transponders, and it would help bring entertainment to approximately 103 million homes in Europe. The ASTRA 1KR was scheduled to replace the ASTRA 1K lost in November 2002 (hence the "KR" designation). The spacecraft's operational life expectancy was at least 15 years.[239]

The ATLAS V/CENTAUR that Lockheed Martin chose for the mission was equipped with one Solid Rocket Booster (SRB), a single-engine CENTAUR upper stage and a four-meter diameter payload fairing. According to the flight scenario, the vehicle would roll into a flight azimuth 84.8 degrees shortly after lift-off. The SRB would be jettisoned about two minutes and 24 seconds into the flight, and the CENTAUR upper stage would fire one minute and 54 seconds later. The payload fairing would be jettisoned about four and one-half minutes after lift-off, but the CENTAUR would continue to provide thrust for an additional thirteen and a half minutes. Following a coasting period, the CENTAUR would begin its second burn at T plus 103 minutes and 56 seconds. The CENTAUR's second burn would end at T plus 105 minutes and 24 seconds, and the payload would separate approximately 108 minutes and 10 seconds after lift-off. On-orbit, the ASTRA 1KR would take up its station at 19.2 degrees East longitude.[240]

Technicians erected the ATLAS V booster on Complex 41 on 1 March 2006, and they added the SRB to the vehicle on 3 March 2006. Engineers mated the upper stage on 8 March, and officials completed the Launch Vehicle Readiness Test (LVRT) between 11 and 21 March 2006. Engineers detected a liquid oxygen leak in the CENTAUR's fill and drain valve during the Wet Dress Rehearsal (WDR) on 29 March, and the vehicle had to be rolled back to the Vertical Integration Facility following de-tanking operations. The valve was replaced, and a proper seal was verified on 3 April 2006. Officials completed the WDR successfully on 4 April, and engineers mated the ASTRA 1KR payload to the launch vehicle on 8 April 2006.[241]

The countdown on 20 April 2006 went well. There were no unplanned holds, and the ATLAS V/CENTAUR lifted off Complex 41 at 2027:00.188Z on the same date. The mission was successful. 45th Space Wing officials dedicated the launch to the memory of Major General Jimmey R. Morrell, the first commander of the 45th Space Wing. (General Morrell passed away on 8 February 2006.) General Morrell served as 9th Space Division Commander from 1 October 1990 through 22 September 1991, and he assumed command of the 45th Space Wing on 23 September 1991. He completed his tour as 45th Space Wing Commander on 29 June 1993.[242]

ATLAS V/CENTAUR Civil Space Operations

ATLAS V/CENTAUR (MARS RECONNAISSANCE ORBITER), 12 August 2005

The object of this ATLAS V/CENTAUR mission was to boost NASA's MARS RECONNAISSANCE ORBITER (MRO) on the first leg of its interplanetary voyage to Mars. According to the flight scenario, the vehicle lifted off Complex 41 on a flight azimuth of 104 degrees. The MRO payload separated from the launch vehicle approximately one hour after launch, and it began its seven-month-long journey to the Red Planet. The MRO was designed to orbit Mars for a full Martian year as it gathered high-resolution imagery and other data on the climate and details of the planet's terrain. The 4,796-pound spacecraft featured six scientific instruments including an imager that could resolve surface features as small as three feet in length. Once the MRO reached Mars, the spacecraft would enter a 300 x 45,000-kilometer polar capture orbit. Aero-braking maneuvers over the next six months would lower the spacecraft's orbit to 255 x 320 kilometers to allow a thorough mapping of the planet. NASA expected the $720 million MRO mission to provide important data on safe landing areas for future Mars missions.[243]

Technicians erected the ATLAS V booster on Complex 41 on 6 May 2005. Engineers mated the upper stage on 10 May, and officials completed the Launch Vehicle Readiness Test (LVRT) between 17 and 28 May 2005. The Wet Dress Rehearsal (WDR) was accomplished on 19 July, and engineers mated the MRO payload to the launch vehicle on 28 July 2005. Technicians detected hydrogen vapor around a hydrogen vent fin connection to a ground umbilical during a preliminary WDR in early July, but the problem was resolved (and closed out during the WDR on 19 July) by replacement of the disconnect assembly and seals associated with the connection. Preparations continued for a launch on 10 August 2005.[244]

An electrical short on 5 August prompted the contractor to replace a faulty battery on 9 August 2005. In the meantime, NASA asked range officials to move the launch from 10 August to 11 August, and the countdown was rescheduled accordingly. Thunderstorm activity early on the 11th delayed the countdown, and there were two unplanned hold during the count. Ultimately, a liquid oxygen tanking software problem prompted officials to scrub the launch at 1251Z on 11 August 2005. Engineers quickly determined the software problem was lightning-induced, and the countdown was recycled for the launch on 12 August 2005.[245]

The countdown on 12 August went well. There were no unplanned holds, and the ATLAS V/CENTAUR lifted off Complex 41 at 1143:00.311Z on the same date. By the end of August, the MARS RECONNAISSANCE ORBITER was more than 1,500,000 miles from Earth on a curving 310-million-mile trajectory to Mars.[246]

ATLAS V/CENTAUR (PLUTO NEW HORIZONS) 19 January 2006

The object of the ATLAS V/CENTAUR launched from Complex 41 was to send NASA's PLUTO NEW HORIZONS spacecraft on the first leg of its journey to Pluto, its moon Charon and two other moons in the Kuiper Belt some 3 billion miles from Earth. The spacecraft was the first of NASA's "New Frontiers" class interplanetary probes, and its journey was expected to take more than nine and a half years. The project (including the launch vehicle) cost approximately $700 million. The Johns Hopkins University Applied Physics Laboratory in Laurel, Maryland, led the development of the 1,054-pound spacecraft, and the probe carried a suite of seven sensors, two 64-gigabit solid-state tape recorders, and a plutonium-fueled radioisotope thermoelectric generator to supply 200 watts of power for all the spacecraft's electrical needs.[247]

The ATLAS V/CENTAUR vehicle chosen for the mission carried five Aerojet Solid Rocket Boosters (SRBs), and Lockheed Martin incorporated a Boeing/ATK Thiokol DELTA II spin table and 15,000-pound Star 48B solid-fueled third stage atop the booster's Pratt & Whitney CENTAUR second stage. The vehicle was the first ATLAS V 500 series to carry five SRBs. Fully fueled, the ATLAS V/CENTAUR for this mission weighed 1,260,000 pounds, and it developed approximately 2.5 million pounds of thrust at lift-off. Flying at 36,000 miles per hour, the spacecraft was expected to reach Jupiter (for a gravity assist) about 13 months after lift-off. If all went well, the probe would reach the vicinity of Pluto in mid-July 2015. Approaching Pluto as close as 6,200 miles and Charon as close as 16,800 miles, the probe was expected to deliver "Landsat quality" images of the surface of Pluto as well as data on the geology of its impact craters. Charon would also be studied by the probe as it orbited Pluto some 12,000 miles away. The spacecraft would begin gathering data on smaller objects in the Kuiper Belt toward the end of 2016.[248]

Technicians erected the ATLAS V booster on Complex 41 on 29 September 2005, and engineers installed the CENTAUR upper stage on 11 October 2005. SRB Stacking began on 19 October, but Hurricane WILMA breached Complex 41's megadoor in early November, and the door touched one of the ATLAS V's SRBs. The SRB was replaced by 4 November, and. all five SRBs were attached to the vehicle by 29 November 2005. Officials completed the Launch Vehicle Readiness Test on 1 December, and the Wet Dress Rehearsal was accomplished on 6 December 2005. An ATLAS RP-1 tank isogrid crack prompted a fleet-wide investigation in November and early December 2005, and official rescheduled the mission from 11 to 17 January 2006 while results were being tabulated. The PLUTO NEW HORIZONS spacecraft arrived at the Kennedy Space Center on 1 December 2005, and engineers mated the payload to the ATLAS V/CENTAUR on 17 December 2005.[249]

NASA cleared flight constraints associated with the isogrid crack on 13 January 2006. The launch did not go off as planned on the 17th, and the countdown on 18 January was scrubbed at 1530Z due to a power failure at the Applied Physics Laboratory Mission Operations Center. Nevertheless, the countdown on the 19th went well, and the ATLAS V/CENTAUR lifted off safely at 1900:00.221Z on 19 January 2006. The launch was reportedly "flawless," and the PLUTO NEW HORIZONS spacecraft was 6 million miles closer to Pluto by 27 January 2006.[250]

TITAN IV/CENTAUR and TITAN IV/IUS Overview

The TITAN IVA and TITAN IVB were upgraded versions of the TITAN 34D, which had been a Defense Department workhorse in the 1980s. For their inaugural flights, the TITAN IVA and the TITAN IVB each employed an Inertial Upper Stage (IUS), but other TITAN IVA and TITAN IVB vehicles employed a CENTAUR upper stage or no upper stage at all. The first TITAN IVA was launched from Cape Canaveral's Complex 41 on 14 June 1989. The TITAN IVB made its debut flight from Complex 40 on 23 February 1997. Each TITAN IVA had a gross weight of approximately 1,900,000 pounds. The TITAN IVB/CENTAUR's maximum weight was about 2,200,000 pounds. A TITAN IVA/CENTAUR could lift 39,100 pounds into low-Earth orbit or place a 10,000-pound payload into geosynchronous orbit. A TITAN IVB/CENTAUR could carry up to 47,800 pounds to low-Earth orbit or place up to 12,700 pounds into geosynchronous orbit. Even when it was equipped with only an IUS, a TITAN IV could place a 5,200-pound payload into geosynchronous orbit. A family of 200-inch diameter payload fairings complemented the launch vehicles' great payload capacity. With an 86-foot-long payload fairing in place, the tallest TITAN IV/CENTAUR stood about 204 feet tall.[251]

Like the TITAN 34D, the TITAN IVA and TITAN IVB core vehicles were configured around three stages: Stage 0, Stage I, and Stage II. A pair of Solid Rocket Motors (SRMs) containing PBAN solid composite propellant was carried on the TITAN IVA as Stage 0. Each SRM weighed about 700,000 pounds, and it consisted of seven segments assembled in a stack 10 feet in diameter and 112.9 feet long. Together, a pair of SRMs developed the vehicle's total initial thrust — about 3,200,000 pounds. A Thrust Vector Control (TVC) system used nitrogen tetroxide (N2O4) to maintain the proper flight path while Stage 0 was firing. The TVC system injected nitrogen tetroxide through 24 electromechanical valves spaced equally around the nozzle of each SRM. The SRMs separated from the TITAN IVA vehicle about two minutes into the flight at an altitude around 164,000 feet.[252]

For the TITAN IVB, Stage 0 consisted of two three-segment Upgraded Solid Rocket Motors (SRMUs) 112.4 feet long and 10.5 feet in diameter. Each SRMU weighed about 771,000 pounds and burned PBAN solid composite propellant. Each SRMU was rated at 1.7 million pounds of thrust at peak vacuum. Alliant Techsystems built the SRMUs, and the motors burned 88 percent Hydroxyl Terminated Polybutadiene. On a typical TITAN IVB/CENTAUR flight, the SRMUs separated 146 seconds after launch at an altitude of 227,000 feet. The SRMUs were equipped with a hydraulic TVC system to keep the vehicle on the proper flight path.[253]

Stages I and II were proportioned the same on both vehicles. Stage I was 86.5 feet long, and Stage II was 32.7 feet long. Both stages were 10 feet in diameter, and they were fueled with a mixture of nitrogen tetroxide and Aerozine 50. Stage I carried about 346,000 pounds of propellant and weighed 373,000 pounds when completely fueled. Stage I developed about 551,000 pounds of thrust from its two liquid turbopump-fed LR87-AJ-11 engines. The engines ignited about 116 seconds into the flight, and they continued to thrust for about 190 seconds before shutting down. In the meantime, the payload fairing was jettisoned between 198 seconds and 236 seconds into the flight at an altitude between 370,000 and 386,000 feet.[254]

Stage II weighed about 86,000 pounds including 77,000 pounds of propellant. Stage II's single LR91-AJ-11 engine developed about 106,000 pounds of thrust. In a typical east coast TITAN IVA or TITAN IVB launch, Stage II ignited between five minutes and five and one-half minutes into the flight at an altitude somewhat higher than 500,000 feet. Stage I was jettisoned less than a second later. Stage II began to burn its own supply of nitrogen tetroxide and Aerozine 50. That burn lasted a little less than four minutes. After Stage II shut down, the CENTAUR and spacecraft (or IUS and spacecraft) separated from the core vehicle about nine seconds later. For a CENTAUR-equipped vehicle, the payload and its CENTAUR upper stage were approximately 1,000 nautical miles downrange at a typical altitude of 692,000 feet at the time of separation.[255]

The CENTAUR stage was 14.2 feet in diameter and 29.5 feet in length. It was equipped with two Pratt & Whitney RL10A-3-3A liquid hydrogen-fueled engines rated at 33,000 pounds of thrust. The CENTAUR's first burn began about ten minutes after lift-off. Depending on the mission, the CENTAUR's first burn varied in length from about three and one-half to five minutes. The first burn injected the spacecraft into a 100-nautical-mile-high parking orbit. Two more CENTAUR burns and several maneuvers were required to place the spacecraft into geosynchronous orbit.[256]

On a typical TITAN IVB/IUS mission, Stage II shut down about eight minutes and 39 seconds after lift-off, and the IUS separated from the TITAN IV about nine seconds later. Following two minutes of internal adjustments, the IUS started a 30-minute thermal roll. A reverse thermal roll was initiated about 43 minutes after lift-off. A guidance hold was acquired about 65 minutes into the flight, and the IUS' first SRM fired about five minutes later. The first SRM continued to thrust for about two and one-half minutes, and the IUS and spacecraft entered a transfer orbit in preparation for the second SRM burn. The second SRM fired about six hours and 30 minutes after the launch, and it continued to thrust for about one minute and 47 seconds. Further adjustments, a reaction control system firing, and a thermal roll were followed by spacecraft separation about six hours and 50 minutes into the flight. In that manner, the spacecraft was delivered to geosynchronous orbit.[257]

TITAN IV space launch operations at the Cape required a tremendous amount of instrumentation support. Optics support included the fixed Intercept Ground Optical Recorder (IGOR) and Contraves coverage from various locations on the range. The Remote Optical Tracking Instrument (ROTI) and the Distant Object Attitude Measuring System (DOAMS) also covered the launch phase for optical purposes. When radars, telemetry systems, communications, command/destruct systems, and a full complement of weather systems were added to the support picture, it was easy to see why a TITAN IV launch was a major effort for the Eastern Range.[258]

TITAN IV Military Space Operations

TITAN IVA/CENTAUR, MILSTAR, 7 February 1994

On 7 February 1994, a TITAN IVA/CENTAUR placed the first MILSTAR spacecraft into orbit.[259] The mission was successful, but preparations for the flight (and other TITAN IVA/CENTAUR flights in 1994) proved lengthy. The TITAN IVA core vehicle (K-10) arrived at the Cape near the end of August 1992. Engineers and technicians assembled, erected, and checked out K-10 in the Vertical Integration Building (VIB) between 2 September 1992 and 22 May 1993. Workers moved the core vehicle to the Solid Motor Assembly Building (SMAB) on 22 May, and technicians mated the booster to its Solid Rocket Motors (SRMs) by 25 May 1993. Following further tests, the launch vehicle was transported to Launch Complex 40 on 3 June 1993. Officials completed a Combined Systems Test (CST) and a Flight Events Demonstration (FED) at the complex by 15 July 1993. Unfortunately, there was a TITAN IVA launch failure at Vandenberg AFB on 2 August 1993,[260] and it compelled Air Force officials to suspend TITAN IV pre-launch operations at the Cape. On 23 September 1993, Air Force officials decided to de-stack, examine, and re-stack K-10's SRMs. Engineers completed those tasks between 28 September and 4 November 1993. Another CST was accomplished on 5 November, and engineers mated the MILSTAR payload to the launch vehicle on 3 December 1993. Payload integration testing continued through 24 January, and officials completed the Launch CST successfully on 25 January 1994.[261]

The MILSTAR mission's first countdown got underway on 5 February 1994, but officials scrubbed it on that date. They recycled the countdown and picked it up again on 7 February. There were no unscheduled holds during that countdown, and the vehicle lifted off successfully at 2147:01.435Z on 7 February 1994. Hailed as the Defense Department's "core command and control communications system for U.S. combat forces well into the next century," the MILSTAR spacecraft was the first of a network of satellites expected to provide a flexible, jam-resistant "switchboard in space" for many years to come.[262]

TITAN IVA/CENTAUR, Classified Payload, 3 May 1994

This mission's TITAN IVA core vehicle (noted hereafter as TITAN IV-7) arrived at Cape Canaveral on 11 May 1990. Engineers assembled and erected the booster in the Vertical Integration Building (VIB) on 23 January 1991. Officials completed the core vehicle's first Combined Systems Test (CST) on 8 April 1991. Following the Solid Rocket Motor (SRM) mate at the Solid Motor Assembly Building (SMAB), a CST retest was accomplished on 31 May 1991. Workers moved the vehicle out to Complex 41 on 14 June, and technicians completed the TITAN IV-7's SRM buildup at the launch pad on 20 June 1991. Engineers mated the CENTAUR upper stage (TC-8) to the launch vehicle on 12 July 1991. After the Baseline CST and Flight Events Demonstration (FED) were accomplished in early August, officials completed a Terminal Countdown Demonstration (TCD) on 29 November 1991. Unfortunately, the mission was held up by a very lengthy investigation into a commercial ATLAS/CENTAUR launch failure on 11 April 1991. The mishap brought the reliability of all CENTAUR upper stages into question. As the investigation continued, the TITAN IV-7 vehicle's upper stage (TC-8) was removed on 6 December 1991, and engineers replaced it with a new upper stage (TC-9) on 7 January 1992. The launch vehicle remained standing at the launch pad for more than a year.[263]

On Martin Marietta's advice, Air Force officials eventually decided to roll the vehicle back to the SMAB for de-stacking operations in early August 1992. Engineers removed TC-9 from the TITAN IV-7 vehicle on 5 August, and they mated TC-9 to the TITAN IV-9 core vehicle on 27 August 1992. (TITAN IV-9 had been awaiting staging to the launch pad from the Solid Motor Assembly & Readiness Facility [SMARF] since 13 May 1992, and TITAN IV-9 replaced TITAN IV-7 at Complex 41 on 12 August 1992.) Another ATLAS/CENTAUR launch failure occurred on 22 August 1992, and, following the official investigation of that mishap, TITAN/CENTAUR upper stages were modified. In the meantime, officials completed TITAN IV-9's Terminal Countdown Demonstration on 12 November 1992, but the vehicle's CENTAUR upper stage (TC-9) had to be removed 12 days later. It was replaced with another CENTAUR (TC-10) on 25 November 1992, but that upper stage was removed on 12 February 1993. Since TITAN IV-9 would have been standing more than a year by the time a *modified* CENTAUR could be cleared for launch, the vehicle was rolled back to the SMARF on 24 March 1993.[264]

In the meantime, the TITAN IV-7 vehicle had two Combined Systems Tests in the VIB on 11 December 1992 and 8 February 1993. Following another SRM mating operation in the SMAB, workers moved TITAN IV-7 out to Complex 41 on 30 March 1993. Technicians completed the SRM buildup at the launch pad on 7 April, and engineers mated the TITAN IV-7 vehicle to TITAN IV-9's recently removed upper stage (TC-10) on 12 April 1993. Officials completed

TITAN IV-7's Baseline CST/FED on 24 May 1993, and engineers mated the spacecraft to the launch vehicle on 11 July 1993. Unfortunately, the TITAN IV launch failure at Vandenberg AFB on 2 August 1993 led to a de-stacking and restacking exercise between 2 December 1993 and 26 February 1994. Using the Mobile Service Tower's crane, technicians de-stacked both of TITAN IV-7's SRMs *with the spacecraft and payload fairing in place*. Another CST was accomplished on 15 March 1994, and the Launch CST was completed on 10 April 1994.[265]

A small (but noteworthy) incident occurred at Complex 41 five days later. Around 1:00 p.m. on 15 April 1994, Range Safety (45 SW/SESL) received a telephone call from Pad Safety officials at Complex 41. The officials called to report the discovery of a small nitrogen tetroxide leak (e.g., about one teaspoon of oxidizer every five minutes) in a long pipe associated with oxidizer loading operations for TITAN IV-7. The leak started around 11:39 a.m. on the 15th, and, with loading operations in progress, Pad Safety asked how Range Safety wanted to handle the situation. A clear zone had already been established around the pad out to a distance of 8,000 feet for loading operations, and the wind (coming from the southeast) was slow enough to preclude exposure to unprotected personnel. Small leaks from hoses, fittings, and quick disconnects were not unusual in TITAN operations. Range Safety concluded there was a hairline crack in the pipe, and they advised Pad Safety to have Martin Marietta dispatch a worker in a SCAPE suit (Self-Contained Atmospheric Protective Ensemble) to wrap the leaking pipe with duct tape. With the duct tape in place, Martin Marietta completed the loading operation on April 15th. The incident was not serious, but it was the first of three significant nitrogen tetroxide leaks at the Cape in 1994.[266]

The first countdown for the TITAN IV-7 mission got underway on 21 April 1994, but officials scrubbed the launch at 1154Z on that date because the 2A1E umbilical was pulled inadvertently from its airborne mating receptacle by a rope attached to it and a nearby Mobile Service Tower platform. The break interrupted power to the vehicle's inertial guidance system, and the launch had to be recycled. The umbilical was subsequently refurbished and reconnected, and the countdown for the second launch attempt got underway on 23 April. There were two problems with the CENTAUR and its ground equipment during the countdown, and one of the range's radars experienced transmitter problems. Those problems were corrected, but officials scrubbed the launch at 1735Z on the 23rd for low clouds and thunderstorms within ten miles of the launch vehicle. They scrubbed the third launch attempt at 1655Z on 26 April after the automatic sequencer failed to arm the vehicle's safety destruct system. Fortunately, the fourth countdown on 3 May 1994 proceeded smoothly, and the TITAN IVA/CENTAUR lifted off Pad 41 at

1555:01.230Z. The flight was nominal, and Air Force officials declared the launch a total success around 1609Z (1209L) on 3 May 1994.[267]

TITAN IVA/CENTAUR, K-9, Classified Payload, 27 August 1994

The TITAN IVA (K-9) core vehicle for this classified mission (noted hereafter as TITAN IV-9) arrived at Cape Canaveral on 4 December 1990. Technicians assembled and erected it on Transporter 1 in the Vertical Integration Building (VIB) on 13 and 14 May 1991. The vehicle's first CENTAUR upper stage (TC-9) arrived on 29 August 1991, and technicians and engineers processed the CENTAUR between 31 August and 13 December 1991. Initially, TC-9 was mated to the TITAN IV-7 vehicle on Complex 41 on 7 January 1992, but it was demated from the vehicle on 5 August 1992 after the Air Force decided to de-stack TITAN IV-7's Solid Rocket Motors (SRMs) at the Solid Motor Assembly Building (SMAB). In the meantime, Martin Marietta completed a Combined Systems Test (CST) on TITAN IV-9 on 27 November 1991. Technicians and engineers built up TITAN IV-9's SRMs between 17 February and 25 April 1992, and they mated them to the core vehicle at the SMAB between 7 and 12 May 1992. Workers rolled the vehicle into the Solid Motor Assembly & Readiness Facility (SMARF) on 13 May 1992 to await staging to the launch pad. TITAN IV-9 core vehicle replaced TITAN IV-7 at Complex 41 on 12 August 1992, and engineers mated TC-9 to TITAN IV-9 on 27 August 1992. Preparations for the TITAN IV-9 launch continued, and officials completed a Terminal Countdown Demonstration (TCD) on the vehicle on 12 November 1992.[268]

Despite this progress, TITAN IV-9's CENTAUR upper stage was demated on 24 November 1992. A new CENTAUR — TC-10 — was installed on TITAN IV-9 on 25 November 1992. Subsequently, that upper stage was removed on 12 February 1993, and engineers mated it to the TITAN IV-7 vehicle on 12 April 1993. Owing to complications arising from an ATLAS/CENTAUR launch failure in August 1992, the TITAN IV-9 vehicle was rolled back to the SMARF on 24 March 1993. Indeed, the various CENTAUR matings, de-matings, and re-matings during this period merely punctuated the contractor's struggle to restack stale launch vehicles and modify CENTAUR upper stages after the ATLAS/CENTAUR mishap investigations in 1991 and 1992. Ultimately, TITAN IV-9's third and final CENTAUR (TC-11) was delivered to the Cape on 13 August 1993. Engineers and technicians processed it between 16 August 1993 and 27 May 1994.[269]

Workers returned the TITAN IV-9 core vehicle to the VIB for modifications and tests on 16 April 1993. Following those actions, the vehicle remained in storage until 6 January 1994. Technicians reapplied power on 6 January 1994, and the core vehicle was moved to the SMAB for SRM buildup and mating on 22 April 1994. Technicians completed the SRM mating on 10 May, and TITAN IV-9 was rolled out to Complex 40 on 11 May 1994. The SRM buildup was completed at the pad on 14 May, and workers moved the vehicle to the SMARF on 26 May for short-term storage. Technicians moved TITAN IV-9 to Complex 41 on 28 May 1994. Engineers mated TC-11 to the launch vehicle on 1 June, and officials completed the baseline CST/FED (Combined Systems Test/Flight Events Demonstration) on 18 June 1994. The TCD was conducted on 26 June, and the Mission Dress Rehearsal was completed on 15 July 1994. Following the Launch CST on 15 August, preparations continued for the first TITAN IV-9 launch attempt on 25 August 1994.[270]

Unfortunately, about 400 gallons of nitrogen tetroxide (N2O4) were spilled in the Oxidizer Holding Area on Complex 41 after a stainless steel flexhose ruptured around 10:20 a.m. on 20 August 1994. Most of the N2O4 was released as a large (600 x 150 x 40-foot) dark reddish brown vapor cloud over the south end of Complex 41. The liquid N2O4 was confined to a bermed concrete containment area, but the vapor chemically burned about 1.5 acres of vegetation surrounding the Oxidizer Holding Area. The people working at Complex 41 evacuated quickly. They were examined by medical personnel and released. The public was in no danger at any time. The Cape Commander assumed control of the incident upon her arrival, and a qualified reconnaissance team was sent to Complex 41 around 12:00 noon (local time) to survey the oxidizer farm. The break in the flexhose was located, and a valve was turned to isolate the ruptured line. The reconnaissance team remained onsite until their air ran low (around 1:15 p.m.). A second team was sent in to clean up and replace the flexhose, and the launch pad was reopened for essential work at 7:45 p.m. on 20 August 1994.[271]

The 5th Space Launch Squadron Commander convened an Operations Review Board (ORB) on 30 August 1994. The ORB's findings echoed Martin Marietta's evaluation of the incident and concluded, "...the incident may have been averted if proper procedure change coordination requirements had been followed." Corrective actions included: 1) insertion of a warning note in the propellant load portion of the procedure, 2) retraining and recertifying the propellant loading and purging technicians and launch conductors before the next propellant loading operation, and 3) Range Safety approval of the plan to recertify the oxidizer system before use.[272]

Following replacement of the flexhose, preparations for the TITAN IV-9 launch on Complex 41 continued. The first launch attempt got underway on 25 August 1994. There were two unscheduled holds during the countdown. The first one lasted 30 minutes to allow personnel to clear the launch pad. The second one required 100 minutes to resolve a toxic fume threat. The countdown continued, but officials ultimately scrubbed the launch at 0906Z on the 25th due to thick cumulus clouds and lightning within 10 miles of the launch complex. The launch was rescheduled for 27 August 1994. Though the launch on the 27th was delayed for two hours due to thunderstorms within five nautical miles of Complex 41, the TITAN IV lifted off successfully at 0858:01.487Z on 27 August 1994. Instrumentation support was very good. The flight profile was nominal, and the launch was successful.[273]

TITAN IVA/IUS, Defense Support Program Satellite, 22 December 1994

The purpose of this mission was to inject a Defense Support Program (DSP) satellite[274] into a proper orbit. The TITAN IVA core vehicle (noted hereafter as TITAN IV-14) arrived at Cape Canaveral on 9 August 1993. Technicians erected the booster between 17 and 22 March 1994. Power was applied to the vehicle on 11 April, and officials accomplished a Combined Systems Test (CST) in the Vertical Integration Building on 19 May 1994. The Solid Rocket Motors (SRMs) were built up in the Solid Motor Assembly Building between 31 May and 21 June. Technicians mated the core vehicle to its SRMs on 25 June 1994. Following TITAN IV-14's arrival at Complex 40 on 5 July 1994, engineers mated the Inertial Upper Stage (IUS) to the vehicle on 14 July 1994. Officials completed a Baseline CST at the launch pad on 1 August, and engineers mated the payload to the launch vehicle on 4 August 1994. Technicians installed the payload fairing on 31 August. Unfortunately, workers discovered cotton fibers inside the payload fairing and on the payload's surface on 3 September 1994. Technicians disassembled the fairing and cleaned the spacecraft and fairing, but more cotton fiber contaminant was found on the fairing and spacecraft on 6 October 1994. Disassembly and cleaning operations were accomplished again, and processing operations resumed on 27 October 1994. Technicians reinstalled the fore and aft payload fairings toward the end of November 1994. A component failed during the Launch CST on 5 December, but the item was replaced. Officials completed the second Launch CST successfully on 8 December 1994.[275]

As launch preparations continued at Complex 40, workers detected a nitrogen tetroxide (N2O4) leak in the Thrust Vector Control (TVC) tank mounted on SRM 1. The leak was first noted around 2:54 a.m. on 15 December 1994. The possibility of a leak was anticipated on 14 December, and the Disaster Control Group was briefed during the night of December 14th. Nevertheless, when the leak finally appeared approximately 300 gallons of N2O4 spilled out before the TVC tank was depressurized to reduce the flow to a minor leak. Officials evacuated the complex. Emergency fire and ambulance teams were notified late, and they arrived about 30 minutes after the evacuation. Medical personnel examined seven people and released them. The Cape Commander was notified, and she responded as on-scene commander at 3:30 a.m. All pad personnel were accounted for by 4:30 a.m., and the 45th Space Wing Vice Commander and 5th

Space Launch Squadron Commander arrived at the scene at 5:30 a.m. The source of the leak was discovered at 12:10 p.m., and the TVC tank and manifold were offloaded by 3:30 p.m. Technicians began purging the TVC tank and manifold with gaseous nitrogen shortly thereafter, and the Cape Commander returned control of the pad to Pad Control personnel. The purge was completed at 11:40 p.m. on 15 December 1994. Technicians refueled and pressurized the tank, and preparations for the launch continued without further incident. An Operations Review Board (ORB) was convened on 11 January 1995 to examine the incident and determine corrective actions.[276]

The countdown on 20 December 1994 experienced some hardware difficulties, but officials terminated the countdown at 0220Z on 21 December 1994 because of cloud thickness and upper level wind factors. Since the vehicle and range were in good shape for a launch, the countdown was recycled quickly. The second launch attempt got underway on 21 December, but officials scrubbed the second countdown at T minus 400 minutes due to projections of bad weather for a launch later in the day. The third launch attempt on 22 December was successful. After an uneventful countdown, the TITAN IVA/IUS vehicle lifted off Pad 40 at 2219:02.068Z on 22 December 1994. Instrumentation support was good, and the vehicle injected its payload into the proper orbit. The launch was a success.[277]

TITAN IVA/CENTAUR, Classified Payload, 14 May 1995

The TITAN IVA core vehicle for this classified mission arrived at Cape Canaveral on 19 July 1994. Technicians assembled and erected the booster in the Vertical Integration Building (VIB) on 20 September 1994, and officials completed the first Combined Systems Test (CST) on the vehicle on 4 January 1995. Following the Solid Rocket Motor (SRM) mate at the Solid Motor Assembly Building (SMAB), workers moved the vehicle to Complex 40 on 11 January 1995. Technicians completed the SRM buildup on 15 January, and engineers mated the CENTAUR upper stage to the launch vehicle on 21 January 1995. Following the Baseline Combined Systems Test (CST) and Flight Events Demonstration (FED) on 13 March, officials completed a Terminal Countdown Demonstration (TCD) on 23 March 1995. Engineers mated the payload to the launch vehicle on 5 April, and the payload fairing was installed on 25 April 1995. Following the Launch CST on 3 May, preparations continued for the launch on 14 May 1995.[278]

Though vehicle processing went smoothly, ground support deficiencies made the TITAN's transport somewhat daunting. About seven weeks before the launch, Complex 40's Mobile Service Tower (MST) "bottomed out" on its rail bed. (The bed had been sinking for years.) The rails were 3.5 inches below grade, and Johnson Controls World Services (the Launch Base

Services contractor) was tasked to remedy the situation. Johnson Controls subcontracted the rail bed emergency work to Hayward Baker, Inc. for $940,000. Hayward Baker's people pumped 2,000 cubic yards of grout under the tracks to raise the rails up to grade. The project was completed seven days before the launch, and it did not impact the launch schedule.[279]

Apart from one unplanned hold of 1 hour and 33 minutes duration, the countdown on 14 May 1995 was uneventful. Officials called the hold for an anomalous CENTAUR signal and a CENTAUR ground equipment failure. Following resolution of those problems, officials resumed the countdown. The TITAN IVA/CENTAUR lifted off Pad 40 at 1345:01.584Z on 14 May 1995. Instrumentation support for the flight went well, and the vehicle's performance, staging events, guidance, and navigation were all within limits. The launch was successful.[280]

TITAN IVA/CENTAUR, Classified Payload, 10 July 1995

The TITAN IVA core vehicle for this classified mission arrived at Cape Canaveral on 30 March 1994. Technicians assembled and erected the booster in the Vertical Integration Building (VIB) on 29 June, and officials completed the first Combined Systems Test (CST) on the vehicle on 2 September 1994. Following the Solid Rocket Motor (SRM) mate at the Solid Motor Assembly Building (SMAB), workers moved the vehicle to Complex 41 on 22 September 1994. The SRM buildup was completed on 28 September, and engineers mated the CENTAUR upper stage to the launch vehicle on 6 October 1994. Officials completed the Baseline CST and Flight Events Demonstration (FED) on 8 November 1994. Unfortunately, a contamination walkdown inspection at the launch pad revealed the CENTAUR Orcon cover had delaminated at the payload fairing hard cover shower head, and the base payload fairing had to be disassembled and cleaned. The base payload fairing and hard cover were subsequently reassembled. A Terminal Countdown Demonstration (TCD) was completed on 3 April, and engineers mated the payload to the vehicle on 29 April 1995. Officials accomplished a Retest CST on 23 May 1995. Following the Launch CST on 30 June, preparations continued for the launch on 10 July 1995.[281]

Before the launch, Complex 41's Mobile Service Tower stopped several times and had to be moved at slow speed to the launch position. Relative humidity in the A-10 Room increased to 69 percent (i.e., 9 percent above limit), but the "user community" accepted the condition and voted to raise the limit. With the exception of the Mobile Service Tower and humidity problems mentioned, all ground support systems performed properly throughout the terminal countdown. There were no unscheduled holds during the countdown, and the TITAN IVA/CENTAUR lifted off Pad 41 at 1238:03.278Z on 10 July 1995. Instrumentation support went well, and the vehicle injected its payload into the desired final orbit.[282]

TITAN IVA/CENTAUR, MILSTAR DFS-2, 6 November 1995

The TITAN IVA core vehicle for the MILSTAR DFS-2 mission arrived at Cape Canaveral on 11 October 1994. The flight was the second in a series of MILSTAR missions, which began in February 1994. Technicians assembled and erected the booster in the Vertical Integration Building (VIB) on 30 January 1995, and officials completed the first Combined Systems Test (CST) on the vehicle on 7 April 1995. Following the Solid Rocket Motor (SRM) mate at the Solid Motor Assembly Building (SMAB), workers moved the launch vehicle to Complex 40 on 30 May 1995. Engineers mated the vehicle to its CENTAUR upper stage on 26 June, and the SRM buildup was completed on 6 July 1995. Unfortunately, Stage II had a number of deficiencies that required correction. Technicians demated the CENTAUR upper stage and Stage II from the core vehicle on 7 August 1995 so engineers could replace the ablative skirt. Engineers re-mated Stage II and the CENTAUR to the launch vehicle on 10 August and 14 August respectively. Officials completed the Flight Events Demonstration and Baseline CST on 30 August and 1 September 1995. A Terminal Countdown Demonstration (TCD) was completed on 8 September 1995. Engineers mated the MILSTAR payload to the launch vehicle on 16 September, and the payload fairing was installed on 16 October 1995. Following the Launch CST on 24 October, preparations continued for the first launch attempt on 5 November 1995.[283]

The countdown for the launch got underway at Complex 40 on 4 November 1995. Unfortunately, officials scrubbed the launch at 0528Z on 5 November 1995 due to an out-of-tolerance 'Red' condition in the Toxic Hazard Corridor (i.e., a safety restriction). Officials recycled the countdown for 6 November. To optimize the launch window, an unscheduled hold was called at T minus 88 minutes to move the T-0 (ignition time) to 0421Z. After calculations were confirmed, the new T-0 was declared at 0228Z, and the countdown resumed at 0243Z. At 0338Z, officials agreed to extend a hold by one minute to accommodate CENTAUR requirements. As a result of out-of-tolerance wind conditions, Lockheed Martin declared a weather hold at 0418Z. The latter part of the launch window looked favorable, and subsequent windsonde (weather) data indicated upper winds were acceptable by 0510Z. The count was picked up at T minus 5 minutes, and the TITAN IVA/CENTAUR lifted off Pad 40 at 0515:01.577Z on 6 November 1995. Eastern Range telemetry support exceeded both Range Safety and user expectations.[284]

The TITAN IVA/CENTAUR injected its MILSTAR payload into the desired orbit on November 6th. All phases of the flight went well. The CENTAUR's flight software navigated and tracked properly during the entire TITAN IV phase of the flight. The mission was successful.[285]

TITAN IVA/CENTAUR, Classified Payload, 24 April 1996

The TITAN IVA core vehicle for this classified mission arrived at Cape Canaveral on 18 December 1991. Following a lengthy storage period (during which the TITAN IV fleet stood down for various reasons), technicians assembled and erected the booster in the Vertical Integration Building (VIB) on 24 July 1995. Officials completed the mission's first Combined Systems Test (CST) on 22 September 1995. Following the Solid Rocket Motor (SRM) mate at the Solid Motor Assembly Building (SMAB), workers moved the launch vehicle to Complex 41 on 28 November 1995. The SRM buildup was completed on 2 December, and engineers mated the CENTAUR upper stage to the launch vehicle on 14 December 1995. Following the Baseline CST and Flight Events Demonstration (FED) on 18 January 1996, officials completed another FED and a Terminal Countdown Demonstration (TCD) on 23 and 29 January 1996 respectively. Following a Launch CST on 30 March 1996 and a Retest CST on 9 April, preparations continued for the launch on 24 April 1996.[286]

There were three unscheduled holds during the countdown on 24 April. A balky environmental shelter door caused the first unscheduled hold at 1635Z, and the hold lasted one hour and 25 minutes. The door refused to open, thereby delaying the Mobile Service Tower's rollback. Officials called the second unscheduled hold at 2105Z for upper level winds out of tolerance. That hold lasted until 2240Z, and it was followed by a safety "No-Go" for toxic indications at 2242Z. Fortunately, the safety No-Go cleared, and the countdown resumed. The TITAN IVA/CENTAUR lifted off Pad 41 at 2337:01.814Z on the 24th. The flight was normal, and no safety actions were required.[287]

TITAN IVA/No Upper Stage, Classified Payload, 3 July 1996

Initially, the TITAN IVA core vehicle designated for this classified mission arrived at Cape Canaveral on 15 July 1988 to support a different launch mission. It was assembled and erected in the Vertical Integration Building (VIB) on 10 September 1988, but it was returned to Martin Marietta in Denver, Colorado on 2 September 1992. Officials eventually selected the vehicle for the July 1996 mission, and it was configured with No Upper Stage (NUS). The core vehicle was delivered to the Cape again on 20 June 1995, and technicians erected the booster in the VIB on 21 December 1995. Officials completed a Combined Systems Test (CST) on the vehicle in the VIB on 1 March 1996. Following the Solid Rocket Motor (SRM) mate at the Solid Motor Assembly Building (SMAB) on 14 March, workers moved the vehicle to Complex 40 on 22 March 1996. The SRM buildup was completed on 28 March, and officials completed the Baseline CST on 21 May 1996. Engineers mated the payload to the launch vehicle on 28 May, and technicians installed the payload fairing on 14 June 1996. Following the Launch CST on 22 June, preparations continued for a countdown on 2 July 1996.[288]

Though bad weather delayed propellant loading operations on 1 July, there were no unscheduled holds once the countdown final got underway the next day. The TITAN IV/NUS lifted off Pad 40 at 0030:01.841Z on 3 July 1996. The vehicle injected its payload into the required orbit successfully.[289]

The launch team dedicated the mission to "all of the Air Force warriors who have lost their lives in Saudi Arabia." The team was referring to the 19 Americans who were killed in the Khobar Towers terrorist bombing near Dhahran, Saudi Arabia, on 25 June 1996.[290] That attack — the worst terrorist attack directed at a U.S. military base since the infamous Marine Barracks attack in Beirut, Lebanon on 23 October 1983 — involved the detonation of 5,000 pounds of explosive in a truck parked against a perimeter fence 85 feet from the nearest dormitory on King Abdul Aziz Air Base. Memorial services were held at Eglin AFB and Patrick AFB on Sunday, 30 June 1996.[291]

TITAN IVB/IUS, DSP Mission, 23 February 1997

The purpose of this TITAN IVB mission was to place a Defense Support Program (DSP) satellite into a predetermined orbit. The TITAN IVB vehicle was equipped with an Inertial Upper Stage (IUS). The mission was a major milestone in the history of the TITAN IV program: it featured the very first launch of a TITAN IVB vehicle anywhere in the world.[292] The TITAN IVB core vehicle arrived at Cape Canaveral on 23 January 1996. Technicians erected the booster in the VIB on 20 February, and officials completed a Combined Systems Test (CST) on the TITAN IVB on 31 July 1996. The SRMUs (upgraded solid rocket motors used exclusively on the TITAN IVB vehicle) were stacked and completed in the Solid Motor Assembly & Readiness Facility (SMARF) by 6 September 1996. Workers moved the launch vehicle to Complex 40 on 13 September, and engineers mated the IUS to the TITAN IVB on 24 September 1996. Officials completed the Baseline CST on 18 October 1996, and engineers mated the payload to the launch vehicle on 7 January 1997. A second CST was accomplished on 9 January, and the Launch CST was completed on 9 February 1997. Flight batteries and ordnance were connected over the following few days, and the vehicle was loaded with propellants and prepared for launch.[293]

The countdown on 23 February 1997 was marred by a two-hour unscheduled hold for technical problems (e.g., Mobile Service Tower closeouts, telemetry RF Link troubleshooting, and IUS checks), but the TITAN IVB/IUS lifted off Pad 40 smoothly at 2020:01.020Z on the same day. The payload was injected into the desired orbit, and the launch was a success.[294]

TITAN IVA/CENTAUR, Classified Payload, 8 November 1997

The purpose of this TITAN IVA/CENTAUR mission was to place a classified Defense Department payload into a predetermined orbit. The TITAN IVA core vehicle arrived at Cape Canaveral on 15 March 1996, and technicians erected it on a transporter on 24 September 1996. Following attachment of the Solid Rocket Motors (SRMs) to the core vehicle at the Solid Motor Assembly Building, workers moved the TITAN IVA to Complex 41 on 13 January 1997. Technicians completed the SRM build-up at the launch pad on 29 January, and engineers mated the CENTAUR upper stage to the launch vehicle on 19 March 1997. Officials accomplished a Baseline Combined Systems Test (CST) on 8 May, and engineers mated the payload to the launch vehicle on 29 May 1997. Another CST was completed on 30 June, and officials completed the launch certification CST on 22 October 1997. Final preparations got underway for the launch on 8 November 1997.[295]

Officials nearly scrubbed the launch on 8 November after an alternate command receiver failed its final checks. Fortunately, the problem was resolved during an unscheduled hold at T minus 5 minutes, and a new launch time was selected. The TITAN IVA/CENTAUR lifted off Pad 41 at 0205:01.646Z on 8 November 1997. The command to separate the spacecraft from the CENTAUR was sent as planned, and separation occurred nominally. The collision and contamination avoidance maneuver (CCAM) and "blowdown" were accomplished, bringing the launch to a successful conclusion.[296]

TITAN IVB/CENTAUR, NRO Payload, 9 May 1998

The purpose of this TITAN IVB/CENTAUR mission was to place a National Reconnaissance Office (NRO) payload into a predetermined orbit from Complex 40. The TITAN IVB core vehicle arrived at Cape Canaveral on 3 October 1997. Technicians erected it on a transporter on 13 November, and workers moved it to the Solid Motor Assembly & Readiness Facility (SMARF) on 21 December 1997. The Upgraded Solid Rocket Motors (SRMUs) for the mission had been stacked and remained waiting at the SMARF since 9 December 1997, so the core vehicle/SRMU mating operation went smoothly on 22 December 1997. Workers transported the launch vehicle to Complex 40 on 30 January 1998, and engineers mated the TITAN IVB to its CENTAUR upper stage on the same day. Following a Baseline Combined Systems Test (CST) on 5 March and a Terminal Countdown Demonstration on 11 March, engineers mated the payload to the launch vehicle on 23 March 1998. The Launch CST was completed on 28 April 1998, and officials certified the vehicle for launch.[297]

The countdown for the launch got underway on 8 May 1998. According to the 45th Range Squadron's post launch report, the countdown on 8-9 May was successful, but it was "fraught with problems." Approximately 150 minutes into the count, a breakaway umbilical line used to pressurize oxidizer in Stage II disconnected itself. A crew of TITAN specialists had to go out to the pad to reattach the line and reestablish pressure. Other difficulties included transient command system problems and the intrusion of a sailboat in the launch contours of the TITAN's flight path. The intrusion led to a brief unplanned hold at T minus 4 minutes, but other problems racked up nearly four hours of 'Red' (Launch No-Go) time during the countdown. Thankfully, the TITAN IVB/CENTAUR lifted off Pad 40 without incident at 0138:00.860Z on 9 May 1998. The TITAN's flight control system and its guidance control unit performed well throughout the flight.[298]

TITAN IVA/CENTAUR, NRO Payload, 12 August 1998

The purpose of this TITAN IVA/CENTAUR mission was to place a National Reconnaissance Office (NRO) payload into a predetermined orbit from Complex 41. The TITAN IVA core vehicle chosen for the mission arrived at the Cape on 24 July 1997. It was an old vehicle, and it would be the last TITAN IVA ever launched. The core was erected on a transporter on 29 January 1998, but it still required significant rework in the Remote Multiplexed Instrumentation System (RMIS) and Wideband Instrumentation System (WIS) wiring before its assembly was completed. Some of this effort involved work on the electrical systems in four compartments where clamps and hardware were replaced and nicked wires were wrapped again. As this work continued, the core vehicle was moved to the Solid Motor Assembly Building on 25 February 1998 where technicians attached its Solid Rocket Motors (SRMs). Workers transported the launch vehicle to Complex 41 on 3 March, and engineers mated the TTIAN IVA to its CENTAUR upper stage on 30 March 1998. The launch had been scheduled for 22 July 1998 previously, but a three-foot-long tear in the CENTAUR's insulation blanket compelled officials to slip the launch to 12 August 1998. The insulation blanket was replaced, and the CENTAUR passed pressure testing on 30 July 1998. Officials completed the Launch Combined System Test (CST) successfully on 31 July 1998, and the Launch Readiness Review (LRR) followed on 10 August 1998. The vehicle was prepared for launch.[299]

The countdown got underway at 2035Z on 11 August 1998. Wind, temperature, and cloud conditions had been well within safety constraints since the CST on 31 July, and those conditions continued. The countdown progressed without incident until the following morning at 0918Z. At that time, a problem surfaced during CENTAUR tanking operations. There was a failure in a ground station electrical relay, and it prompted engineers to discontinue automatic tanking and finish up the operation manually. The manual procedure delayed the launch approximately 88 minutes, but tanking was completed successfully at 1049Z. The TITAN IVA/CENTAUR lifted off Pad 41 at 1130:01.356Z on 12 August 1998.[300]

At T plus 39.463 seconds, an alarm issued — input voltage to the Missile Guidance Computer (MGC) had fallen, causing the start of a "power down" sequence for the MGC. At T plus 39.650 seconds, the MGC recovered power and reinitiated the timing reference signal to the IMU, but the IMU came back with a false indication, prompting the MGC to command the TITAN IVA into a full pitch down with a yaw to the right at T plus 39.818 seconds. In less than two seconds, the disaster ensued. When the TITAN IVA's pitch drifted 13 degrees off the planned trajectory, the vehicle started coming apart. At T plus 41.545 seconds, SRM #1 separated from the core vehicle and triggered the Inadvertent Separation Destruct System (ISDS). The explosion destroyed the core, and SRM #2 was destroyed by its ISDS at T plus 41.709 seconds. At T plus 45.529 seconds, flight controllers sent command destruct signals to the vehicle to ensure its destruction. Based on telemetry, the TITAN IVA reached an altitude of 17,047 feet when it exploded. It was 4,422 feet downrange, and it was traveling at 1,007 feet per second when it disintegrated. Several thousand pieces of propellant and vehicle fragments scattered themselves over a five-mile by three-mile area, but no debris impacted land. No one was in any danger from the mishap, and a 20-knot wind took the cloud of unspent rocket propellant harmlessly out to sea.[301]

According to Brigadier General F. Randall Starbuck, the Wing's response to the accident was "executed flawlessly." Public Affairs released information on the launch failure to the local, national, and international media via press conferences and news releases. The releases were timely, and media response was "balanced and positive in nature." Within 20 minutes of the mishap, the Launch Disaster Control Group (LDCG) was cleared to begin searching the beach for debris. Half an hour later, the LDCG confirmed no debris had landed on the Cape. The LDCG contacted the Coast Guard around 1200Z to close the debris area to marine traffic and search for floating debris. The LDCG also asked NASA for its assistance, and NASA's *Liberty Star* and *Freedom Star* (operated by United Space Alliance) departed at 1030Z on the 13th to help recover floating debris. The U.S. Navy Salvage Group was contacted four days later. Over the next few weeks, Navy personnel mapped the debris field, and they recovered debris lying on the ocean floor 12 to 50 feet below the surface. Approximately 30 percent of the TITAN IVA was salvaged from the ocean, and recovery teams also collected debris that washed ashore for several days after the mishap. The recovery effort was terminated on 15 October 1998.[302]

The Air Force Space Command (AFSPC) Commander appointed his Director of Operations, Major General Robert C. Hinson, to serve as President of the TITAN IVA Accident Investigation Board for the mishap. Colonel Daniel A. Dansro was appointed as the Board's vice president, and eight other officials were selected to serve as technical advisors. In the course of a detailed analysis of telemetry data associated with the accident, the Board learned that at least 30 electrical shorts occurred within various parts of the TITAN IVA vehicle during its unfortunate flight. To determine the cause of the shorts and their ultimate impact, the Board ordered vibration tests, controlled experiments, computer simulations, failure analyses, manufacturer tests, pedigree reviews, and modeling. In January 1999, the Board released the results of its investigation. It was the Board's opinion that "we cannot state with clear and convincing evidence that the mishap was the result of poor workmanship, handling, transportation, or simple carelessness in the entry or exit of a [wiring] compartment." Nevertheless, it was abundantly clear that factory and pre-launch inspections failed to detect the flaw that led to the accident on August 12th. In light of the mishap, the Air Force ordered additional inspections of TITAN IVB and TITAN II vehicles already in the pipeline for future missions. Sure enough, several damaged or "open" wires were discovered with torn insulation or exposed conductors. Lockheed Martin would have to place much greater emphasis on this problem area to avoid similar mishaps in the future.[303]

TITAN IVB/IUS, Defense Support Program 19 Spacecraft (DSP-19), 9 April 1999

The purpose of this TITAN IVB/IUS mission was to place the Defense Support Program 19 spacecraft (DSP-19) into a predetermined orbit from Complex 41. Defense Support Program spacecraft were designed to provide early warning of possible ballistic missile attack on the United States and/or its military forces. In accordance with the launch schedule, the TITAN IVB/IUS vehicle lifted off Pad 41 on 9 April 1999. Unfortunately, the flight did not go according to plan.[304]

Workers moved the TITAN IVB core vehicle out of the Vertical Integration Building (VIB) on 4 September 1998. Engineers mated the IUS to the TITAN IVB on 19 November 1998, and officials complete a Baseline Combined Systems Test (CST) on 11 January 1999. Engineers mated the DSP payload to the launch vehicle at Pad 41 on 17 February, and technicians installed the payload fairing on 13 March 1999. Following the Launch CST on 29 March, preparations continued for the launch on 9 April 1999.[305]

During the countdown on 9 April 1999, the Universal Environmental Shelter Door opened very slowly, and tower rollback was delayed about 90 minutes. As a result of this delay, the T minus 5 minutes hold was extended to a total of one hour and 28 minutes. Though several outages were noted during the countdown, the range remained 'Green' (ready to support). There were no more unplanned holds, and the TITAN IVB lifted off Pad 41 at 1701:00.008Z on 9 April 1999.[306]

Launch operations and the TITAN IVB's performance were normal, but things started going wrong when the first and second stages of the IUS were prompted to separate about six and one-half hours into the flight. Put simply, one of the connectors in the IUS failed to disconnect properly, and it destabilized the IUS second stage and the spacecraft. When mission sequencing commanded spacecraft separation later in the flight, the DSP spacecraft separated cleanly from the IUS *but it was placed in a useless orbit*. Ground controllers tried to stabilize the DSP and raise its orbit to the point it would be useful, but there wasn't enough fuel to save the mission. After making sure the spacecraft would not pose a hazard to other satellites, Air Force Space Command placed the DSP-19 in "on-orbit storage" on 21 May 1999.[307]

TITAN IVB/CENTAUR, MILSTAR (Flight 3), 30 April 1999

The purpose of this TITAN IVB/CENTAUR mission was to place a MILSTAR (Flight 3) military communications spacecraft into a predetermined orbit from Complex 40. According to the flight scenario, the TITAN IVB/CENTAUR was scheduled to lift off Pad 40 on 30 April 1999. The TITAN IVB core vehicle chosen for the mission was moved out of the Vertical Integration Building (VIB) on 22 September 1998. Engineers mated the CENTAUR upper stage to the TITAN IVB on 16 October 1998, and officials completed the Baseline Combined Systems Test (CST) on 20 January 1999. Following the Terminal Countdown Demonstration (TCD) on 28 January, engineers mated the MILSTAR payload to the launch vehicle on 15 February 1999. Technicians installed the payload fairing on 18 March, and the Launch CST was completed on 19 April 1999. Preparations continued for the launch on 30 April 1999.[308]

During the countdown on 30 April 1999, officials extended the built-in hold at T minus 180 minutes to one hour and 34 minutes to complete vehicle processing delayed by rain and lightning. At 1307Z, a new "T-0" (ignition time) was set at 1630Z. The count resumed at 1320Z, and the TITAN IVB/CENTAUR lifted off Pad 40 at 1630:00.976Z on 30 April 1999. The flight went well initially, but an official accident investigation board chaired by Colonel J. Gregory Pavlovich later determined that the CENTAUR's software was corrupted. Ground controllers placed the MILSTAR in a slightly higher orbit (e.g., 2,781 miles) and turned it off. On 12 May 1999, Air Force officials declared the MILSTAR "dead" in orbit.[309]

As unmanned mission failures went, the MILSTAR mishap was certainly one of the costliest in the Cape's history. By early September 1999, an independent review team led by A. Thomas Young (former President of Martin Marietta) determined that Lockheed Martin's launch and spacecraft failures over the previous two years were attributable to: 1) an over-emphasis on cost-cutting as part of the corporation's "faster, better, cheaper" concept, 2) loss of experienced personnel, and 3) poor quality assurance measures. The review team recommended Lockheed Martin develop and implement a TITAN IV "fly-out plan" to retain and reward employees with critical skills and provide better training, mentoring, and succession planning. The "faster, better, cheaper" concept proved to be faster, worse, and disastrously expensive. It was condemned for disrupting proven procedures and, ultimately, costing Lockheed Martin much of its credibility with customers. It remained to be seen if Lockheed Martin would recover its reputation on the heels of three of the most expensive unmanned space flight mishaps in Cape Canaveral's history.[310]

TITAN IVB/IUS, Defense Support Program (DSP-20) Payload, 8 May 2000

The purpose of this TITAN IVB/IUS mission was to place the Defense Support Program 20 spacecraft (DSP-20) into a predetermined orbit from Complex 40. Technicians moved the TITAN IVB core vehicle out of the Vertical Integration Building (VIB) on 22 July 1999. Engineers mated the IUS to the TITAN IVB on 24 September, and officials completed a Baseline Combined Systems Test (CST) on 21 October 1999. Engineers mated the DSP payload to the launch vehicle on 23 November 1999, and technicians installed the payload fairing on 19 January 2000. The Launch CST was scheduled to take place on 29 March 2000, but some cracked solder joints were discovered in early March, so officials completed another Launch CST on 27 April 2000 after repairs were completed. Preparations continued for the launch on 8 May 2000.[311]

During the countdown on 8 May, some data landlines started taking hits at 0833Z. As a result, one of the range's command links was considered Non-Mission Capable (NMC), and the range went 'Red' for safety purposes. The link was restored at 0933Z, but it started taking hits again about three hours later. Once again, the range went 'Red' at 1230Z for the problem. Following one more round of interference around 1539Z, officials declared the range 'Green' (i.e., ready for launch) at 1550Z. A new "T-0" (ignition time) of 1601Z had been coordinated earlier to allow Lockheed time to reinstall an item on the launch vehicle. The reinstallation of the item and the command link problem combined to require a T minus 5-minute hold of 151 minutes. Nevertheless, the TITAN IVB/IUS eventually lifted off Pad 40 at 1601:00.750Z on 8 May 2000. The flight was successful, and the spacecraft separated from the IUS as planned about seven hours later. Following the Cape's string of three TITAN IV failures in 1998 and 1999, the success of this launch was welcome news indeed.[312]

TITAN IVB/CENTAUR, MILSTAR (Flight 4), 27 February 2001

The purpose of this TITAN IVB/CENTAUR mission was to place the MILSTAR (Flight 4) military communications spacecraft into a predetermined orbit from Complex 40. The MILSTAR was designed to provide jam-resistant, high-data-rate, digital communications to command centers and military forces in the field. Military planners had been looking forward to the MILSTAR (Flight 4) mission ever since the MILSTAR (Flight 3) payload was lost in a flight mishap on 30 April 1999.[313]

The launch was scheduled for 17 October 2000, but officials delayed it for a variety of reasons until February 2001. The launch contractor moved the TITAN IVB core vehicle out of the Vertical Integration Building (VIB) to the Solid Motor Assembly & Readiness Facility (SMARF) on 18 May 2000. Unfortunately, the contractor had to delay movement of the vehicle to the launch pad until 19 June 2000 because of a problem with the Mobile Service Tower (MST) drive system.[314] The Baseline Combined Systems Test (CST) was scheduled for 4 August, but range officials agreed to reschedule it for 8 August to accommodate additional delays in processing. The contractor completed installation of the payload fairing on 13 January 2001, and officials completed the Launch CST on 24 January 2001. Officials and contractors looked forward to the launch, which was tentatively rescheduled for 4 February 2001.[315]

In the meantime, a suspect Inertial Navigation Unit (INU) on an ATLAS vehicle had officials worried about a similar unit on the TITAN IVB/CENTAUR. Until the ATLAS INU was "exonerated," processing had to be suspended on the MILSTAR (Flight 4) mission. The ATLAS INU was finally cleared for use on 17 February, but technicians ran into a problem at the pump station that took a day to fix.[316] A problem with the command/destruct system surfaced during final electrical tests and ordnance tests. Resolution of that problem took three days. With that final delay behind them, officials pressed on with the MILSTAR launch on 27 February 2001.[317]

The range experienced "numerous minor instrumentation problems" during the countdown on 27 February, but only three items of a more serious nature deserved comment in the post launch report. The first item involved one of the Cape's radars. Officials declared the radar Non-Mission Capable (NMC) several times during the countdown, notably during the terminal count. The second item involved a launch agency delay. The launch agency was behind schedule, so it declared a "No-Go" at 1545Z. As the launch agency completed its checklist during the extended hold, it experienced a problem with a guidance control unit, and that item delayed the launch, too. The guidance control unit anomaly was corrected at 1731Z, and the launch agency coordinated a new T-0 (ignition time) of 2120Z. Officials picked up the count again at 1810Z. The third notable incident during the countdown involved data landlines. The range went 'Red' at 1842Z after more than four hours of problems involving commercial circuits carrying data. Fortunately, a waiver was received to reroute the data via satellite, and Range Safety was "Go for Launch." The countdown proceeded to the new T-0 (mentioned earlier). The TITAN IVB/CENTAUR lifted off Pad 40 at 2120:00.316Z on 27 February 2001. The mission was successful.[318]

TITAN IVB/IUS, Defense Support Program 21 (DSP-21) Spacecraft, 6 August 2001

The purpose of this TITAN IVB/IUS mission was to place the Defense Support Program 21 (DSP-21) spacecraft in a predetermined orbit from Pad 40. According to *Aviation Week & Space Technology*, the $256 million satellite weighed approximately 5,200 pounds. It was designed to join America's constellation of DSP spacecraft to provide early warning against missile attacks on the United States and/or its military forces.[319]

The DSP-21 payload arrived at the Cape on 8 March 2001. Due to a delay in IUS simulator testing, officials postponed the upper stage mating operation from 12 April to 16 April 2001. It was completed on 23 April 2001, but technicians subsequently discovered some of the connector holes between the core vehicle and the IUS were misaligned. On 2 May, Lockheed Martin officials recommended de-mating the IUS, re-drilling the holes, and re-mating the IUS. The Air Force, Range Safety, and the appropriate contractors approved the recommendation on 4 May, and corrective action solved the mating problem shortly thereafter. The Baseline Combined Systems Test (CST) was scheduled for 21 May, but faulty test equipment delayed the test until 23 May 2001. Engineers mated the spacecraft to the vehicle on 4 June, but technicians found an improperly milled IUS interstage connector that had to be sent back to the factory for correction. The IUS interstage connector was reworked and reinstalled in the vehicle by 25 June 2001. In the meantime, the IUS Flight Termination System (FTS) batteries failed a test, and a Shuttle IUS avionics battery had to be substituted for the FTS battery in early July 2001. Officials approved the avionics battery to fly on 6 July 2001. Technicians completed installation of the DSP-21 payload fairing on 6 July, and officials completed the Launch Combined Systems Test on 16 July 2001. As if all those problems and delays were not enough, an Upgraded Solid Rocket Motor (SRMU) Power Switching Assembly (PSA) failed voltage requirements during factory testing on 25 July. Fortunately, the failure was attributed to the test set power supply, and it had no impact on the flight worthiness of the SRMUs. Officials decided to delay the launch (scheduled for 27

July) until 6 August 2001 after the vehicle's Guidance Control Unit (GCU) was "found out of family" during three navigation tests on 26 July. The GCU was finally cleared to fly by the end of July, and officials also cleared the launch vehicle's SRMU Power Switching Assembly *once again* after a different PSA failed to pass voltage requirements at the factory on 3 August 2001. At long last, final preparations got underway for the launch on 6 August 2001.[320]

Despite all the problems noted earlier, only one launch attempt was needed to send the DSP-21 into space. As with most countdowns, there were many minor instrumentation problems on 6 August, but only three items were really significant. The range was 'Red' for very bad weather between 2008Z on 5 August and 0248Z on 6 August 2001. The chance of a launch seemed minimal at the beginning of the period, but the weather began to improve by 0126Z on the 6th. Weather was clear for launch at 0248Z. In the meantime, the Pad Safety Display was declared Non-Mission Capable (NMC) at 0146Z for a loss of data. That anomaly could have halted the launch, but technicians isolated its source about 90 minutes later, and Mission Flight Control deemed the anomaly too isolated to impact the launch. The range remained 'Green,' and the countdown continued. The dome drive at one of the range's camera sites went NMC at 0705Z, just minutes before the launch. The Kineto Tracking Mount (KTM) tracker at another site went NMC at the same time. Since both deficiencies affected the user but not Range Safety, it was the user's call on whether or not to continue the countdown. The user decided to "Go for Launch". Without further ado, the TITAN IVB/IUS lifted off Pad 40 at 0728:00.091Z on 6 August 2001. The launch was successful.[321]

TITAN IVB/CENTAUR, MILSTAR (Flight 5) 16 January 2002

The purpose of this TITAN IVB/CENTAUR mission was to place the MILSTAR (Flight 5) military communications spacecraft into a predetermined orbit from Complex 40. Like the MILSTAR (Flight 4) spacecraft launched from the Cape in February 2001, the MILSTAR (Flight 5) was one of the world's largest communications spacecraft (e.g., 51 x 116 feet with solar panels extended). It weighed approximately five tons, and it was designed to provide non-jammable, high-data-rate, digital communications to command centers and military forces around the world.[322]

The launch was scheduled for 15 January 2002, and — despite a host of hardware problems — Lockheed Martin came very close to meeting that schedule. The launch contractor moved the TITAN IVB core vehicle out of the Vertical Integration Building (VIB) to the Solid Motor Assembly & Readiness Facility (SMARF) on 22 August 2001. Due to severe weather, the CENTAUR mate was delayed from 14 to 18 September 2001, and the Baseline Combined

Systems Test (CST) slipped from 13 October to 20 October 2001. In the meantime, the launch vehicle's payload fairing connectors hadn't been qualified for launch when the payload fairing was moved out to the launch pad on 1 October. Testing to ensure the connectors' flight worthiness was completed in November, and the connectors were qualified for flight in early December 2001. Other problems included CENTAUR 250-amp hour battery test failures and CENTAUR engine vibrations. Two new batteries were acquired and configured for launch with one of the old batteries, and officials closed out that item on 16 November 2001. Following Lockheed Martin's review and analysis of vibration data, officials cleared the launch vehicle's CENTAUR engine (TC-19) for flight on 9 October 2001. No sooner had that problem been resolved than a problem surfaced involving the CENTAUR's electrical umbilical door mounting holes. The contractor decided to replace the door rather than re-drill it, so technicians removed the TC-19 door and replaced it with the umbilical door from TC-23. Engineers eliminated a software "glitch" in the Programmable Aerospace Ground Equipment (PAGE) chassis on 19 October 2001, but an umbilical cable power problem surfaced on 19 October to keep troubleshooters busy until 16 November 2001. A software coding error that caused test aborts on 29 October was resolved during the spacecraft/upper stage CST on 28 November 2001. Technicians had to replace two Flight Termination System (FTS) antennas on 17 and 18 December 2001 after an MA Communications FTS antenna test failure on 13 December. Another manufacturer, Watkins-Johnson, provided the flight-approved replacements.[323]

Taken together, the aforementioned problems delayed processing in several areas. Officials rescheduled the Baseline CST from 20 October to 30 October 2001. The Terminal Countdown Demonstration was completed on 6 November (nine days later than planned), and engineers mated the spacecraft to the launch vehicle on 18 November 2001 (11 days later than planned). Nevertheless, technicians installed the payload fairing on 17 December 2001, and officials completed the Launch CST as planned on 5 January 2002. Finally, a 5-amp battery failed two load tests on 11 January 2002, but it was replaced within a day. The vehicle was prepared for launch as scheduled. With that last problem behind them, officials pressed on with the MILSTAR (Flight 5) countdown on 15 January 2002.[324]

The range experienced "numerous minor instrumentation problems" during the countdown on 15 January, but only four items of a more serious nature deserve comment. The first item involved weather. The range was 'Red' between 0800Z and 1811Z for thick clouds. In addition, the toxics restriction maximum was exceeded from 1641Z until 1914Z, keeping the range 'Red' until the close of that period. As the wind changed over the course of the next two hours and forty minutes, the range went 'Red' again for toxics, but the condition finally cleared at 2155Z. In the meantime the second item surfaced when the range lost data reception on a communications link. Technicians readjusted the bit rate and re-established reception at 1821Z. At 1903Z, one range station became Non-Mission Capable (NMC) for dropouts/hits on all radar and command communications links. Fortunately the problem cleared on all lines at by 2007Z, and the station was declared Fully Mission Capable (FMC) thereafter. The fourth item concerned a range camera site that experienced data dropouts earlier in the day. The site lost High Density Data (HDD) outputs late in the count. To make up for that deficiency, the instrument on the site was repositioned to view the launch off the pad.[325]

Those items aside, the TITAN IVB/CENTAUR lifted off Pad 40 at 0030:00.674Z on 16 January 2002. The launch was successful. The payload separated from the CENTAUR upper stage properly, and the MILSTAR spacecraft entered the target orbit about six and one-half hours after liftoff.[326]

TITAN IVB/CENTAUR, MILSTAR (Flight 6), 8 April 2003

The purpose of this TITAN IVB/CENTAUR mission was to place the MILSTAR (Flight 6) military communications spacecraft into orbit from Complex 40. Like the MILSTAR (Flight 5) spacecraft launched on 16 January 2002, the MILSTAR (Flight 6) was designed to provide non-jammable, high-data-rate, digital communications to command centers and military forces in the field. The MILSTAR spacecraft weighed approximately five tons, and it measured 51 feet by 116 feet (with solar panels extended) on orbit. It had an operational life expectancy of 10 years.[327]

The mission got off to a rather unusual start when authorities decided to swap the MILSTAR (Flight 6) payload and its CENTAUR upper stage for the NROL-19 payload/upper stage on Complex 40 in early August 2002. Officials decided to swap the payloads (but not the core vehicle) after the National Reconnaissance Office encountered spacecraft difficulties precluding a timely launch. Initially, the MILSTAR (Flight 6) mission was scheduled to launch on 21 January 2003, but unresolved hardware problems from the earlier NROL-19 effort continued to linger as the MILSTAR-6 schedule progressed. In this instance, the launch vehicle's MA/COM II Flight Termination System (FTS) antennas still required RF link analysis before officials could approve the FTS for launch.[328] Additionally, the TITAN IVB's five amp-hour batteries failed qualification testing back in January 2002. A recent shipment of new batteries had to be inspected and activated successfully before that problem was resolved.[329] The vehicle's 250 amp-hour CENTAUR batteries were also somewhat suspect, and failure analysis was needed to confirm that the current stock of batteries was acceptable for flight.[330] On 7 October 2002, an

RL10B-2 development engine (i.e., a CENTAUR engine at the factory) experienced a "low chamber pressure abort" during testing. Since the MILSTAR-6's upper stage had already been mated to its launch vehicle on 17 September 2002, this latest anomaly was the cause of considerable concern. Fortunately, the cause of the RL10B-2 failure was linked to turbine rotor blades manufactured in 2002. Since the RL10A blades on the MILSTAR-6's CENTAUR were manufactured in 1997-98, the anomaly had no impact on the MILSTAR-6 schedule. The System Program Office (SPO) closed out the issue on 15 January 2003.[331]

Three other processing anomalies posed problems for the mission as well. The first of them surfaced during preparations for the Baseline Combined Systems Test (CST) on 21 October 2002. On that date, technicians discovered the CENTAUR's bulkhead GN2 (Gaseous Nitrogen) purge line was crimped. They had to repair the line and confirm its proper function before the Terminal Countdown Demonstration (TCD) could be completed successfully on 2 November 2002. Processing continued without further incident until 8 November 2002. On that date, the trailer transporting the MILSTAR (Flight 6) to Complex 40 disconnected from its towing vehicle as the latter was making a left turn. The payload was apparently unharmed, but a thorough review was required to confirm the status of the spacecraft. The 3rd Space Launch Squadron established an operational review board to investigate the trailer separation, and Detachment 8, Space and Missile Systems Center took measures to ensure the spacecraft had not been damaged in the incident. The spacecraft mate was delayed about two weeks as a result. Officials approved the spacecraft for further operations in late November, and the spacecraft was mated to the vehicle successfully on 25 November 2002. Finally, on 14 January 2003, modeling analysis predicted the CENTAUR's RL10A-4-1A engines might exhibit "differential shutdown" (i.e., less than precisely simultaneous cut-off). The launch date was placed in the "indefinite" category as a consequence, but officials tentatively rescheduled the launch for 2 February 2003 about nine days later.[332]

Software engineers created a successful workaround solution to solve the CENTAUR's differential shutdown problem in late January 2003. Nevertheless, CENTAUR chamber pressure fluctuations surfaced on 21 January 2003, and that phenomenon delayed the launch yet again. As engineers wrestled with the anomaly, officials rescheduled the launch for 4 February. Unfortunately, the problem proved too difficult to resolve quickly. The Space and Missile Systems Center requested a new launch date of 6 April 2003, and 45th Range Squadron approved that request on 20 February 2003.[333]

Following a successful Launch CST on 20 March 2003, preparations were concluded for a first launch attempt on 6 April 2003. Unfortunately, officials had to scrub the countdown on 6 April due to CENTAUR problems, and they recycled the launch for 8 April 2003. The countdown on the 8th went well. According to the 45th Space Wing Command Post message confirming the launch, the TITAN IVB/CENTAUR lifted of Complex 40 successfully at 1343:00.557Z on 8 April 2003. Instrumentation problems during the launch were generally minor, but flight analysts determined that one radar suffered several periods of degraded track due to excessive elevation bias. Fortunately the "large number of radar sources" covering the flight minimized the impact of that anomaly. The MILSTAR spacecraft entered the proper orbit at approximately 2017Z on 8 April 2003. The mission was successful.[334]

TITAN IVB/CENTAUR, NROL-19, 9 September 2003

The purpose of this TITAN IVB/CENTAUR mission was to place a National Reconnaissance Office (NRO) payload into a predetermined orbit from Complex 40. The TITAN IVB core vehicle *initially* chosen for the flight was rolled out to Complex 40 on 28 January 2002, but the launch vehicle's Eagle-Picher five amp-hour batteries failed qualification testing. Ultimately, the battery problem was resolved in late December 2002 after a shipment of new batteries passed qualification testing. In the meantime, several other notable problems arose to delay the NROL-19 mission for more than a year.[335]

On 4 February 2002, an X-Ray inspection of the CENTAUR's rocket engine module revealed a possible blockage. Consequently, shipment of the TC-20 upper stage was delayed until officials dismissed the blockage as a phantom issue on February 8th. Airlift mechanical problems delayed the CENTAUR's arrival at the Cape until 13 February, and the upper stage mate was delayed from 19 through 24 February due to high winds and other weather concerns. Engineers mated the CENTAUR to the launch vehicle on 25 February 2002. In the meantime, an "unauthorized venting of hydrazine" on 14 February and a propane leak on 24 February prompted partial evacuations of Complex 40 on those dates. False fire alarms on the 26th forced evacuation of the Mobile Service Tower, and that problem kept the crew away from the site until 1 March 2002. During payload fairing operations on 20 March, a socket wrench extension fell about 20 feet and dented the thermal Mylar barrier around the CENTAUR upper stage. Fortunately, the CENTAUR remained untouched in the incident, and "there was no structural concern for [the] flight." The Baseline Combined Systems Test (CST) on 10 April 2002 was suspended due to two unplanned CST aborts. The Baseline CST was attempted again on 15 April. The first attempt on that date was aborted, but the second attempt was successful. The Baseline CST was completed, but post-test data revealed an out-of-family measurement on one of the Solid Rocket Motors' High Energy Firing Units (HEFU). That problem would have to be resolved later on. In the meantime, the Terminal Countdown Demonstration (TCD) slipped from 17 to 22 April 2002, and the TCD on the 22nd also experienced a problem — the CENTAUR #2 engine's liquid oxygen inlet turbo pump was out of limits. An engineering review board was held on 9 May 2002, and a special tanking test on 11 June 2002 finally resolved the issue.[336]

Unfortunately, the mission was beset with more problems in late May, June, and July 2002. In late May 2002, Pratt & Whitney's technicians detected a burr on the thrust control valve restrictor orifice of a CENTAUR in production at the company's factory. Because Pratt & Whitney's people hadn't discovered the burr on any earlier inspections of the upper stage, all restrictor assemblies on all CENTAUR upper stages were considered suspect. As a precaution against the possibility of a flight failure, Pratt & Whitney fabricated new restrictors for the upcoming NROL-19 mission and shipped them to the Cape for "immediate installation" on the launch vehicle. Engineers installed the new restrictors successfully on 24 May 2002. For the moment it appeared the mission might be launched in early August 2002, but the customer began experiencing "spacecraft issues" a few weeks later. Consequently, the customer requested a 4 December 2002 launch date, and that request was approved on 26 June 2002. The situation did not improve during the next two weeks, so the launch slipped into the "indefinite" category as officials discussed the possibility of replacing NROL-19 with another mission. By mid-July, it was certain that the NRO could not support a December 2002 launch date. Officials decided to swap out the NROL-19 and remove its CENTAUR upper stage in early August 2002. The core vehicle remained on Complex 40 to support the MILSTAR (Flight 6) mission that was launched on 8 April 2003.[337]

Following the MILSTAR (Flight 6) launch, processing for the NROL-19 mission resumed. The scheduled dates for the upper stage mate, Baseline CST, and Terminal Countdown Demonstration changed several times in April and May, but technicians managed to transport the new core vehicle to Complex 40 just three days later than planned (24 April versus 21 April 2003). As operations continued at the launch pad, technicians detected a hydraulic fluid leak in one of the Universal Environmental Shelter's door actuators. The defective seal responsible for the leak prompted repairs on several actuators before officials re-accepted the door for operations on 10 May 2003. Engineers attempted an upper stage mate on 16 May 2003, but the 25-ton crane malfunctioned during the operation. Technicians had no choice but to return the CENTAUR to its protective trailer and move it back to the Vertical Integration Building (VIB) until the crane was repaired. The malfunction was traced to a power supply in the crane's operational interface console. The power supply was replaced, and the crane was proof-loaded and returned to service by 18 May 2003. The CENTAUR was mated to the vehicle on 19 May, and the core vehicle was powered up by 27 May 2003.[338]

Following lightning strikes within five nautical miles of Complex 40 on 4 June 2003, a brief search of the immediate area revealed "visible damage" to the complex's #2 Uninterruptible Power Supply (UPS). Officials conducted a thorough search of the complex to detect additional damage, and they concluded the matter with the completion of a successful Baseline CST on 13 June 2003. In the meantime, the Programmable Aerospace Ground Equipment (PAGE) aborted a sequence system test on 11 June due to a support equipment error. Troubleshooting corrected the problem, and engineers completed the sequence system test successfully the next day. Officials completed the Terminal Countdown Demonstration on 21 June 2003, and preparations continued for the Launch CST tentatively scheduled for early August. On 22 July 2003, engineers detected some excessive voltage readings from SRMU #2 during flight power monitor testing. Further investigation by the engineers uncovered similar readings during SRMU #1 'power up' operations. Lockheed Martin suspected the cause was moisture in the SRMU raceway cables, and they isolated the problem to moisture on the W117 cable. On 25 July 2003, an engineering review board recommended monitoring "susceptible circuits" for any voltage anomalies until the Launch CST was completed. Engineers completed the Launch CST successfully on 7 August 2003.[339]

Officials expected to launch the NROL-19 mission on 18 August 2003, but Stage II fueling operations were halted on 11 August due to weather concerns. As fueling operations resumed on 12 August, a pump failed around 6:00 p.m., releasing approximately 40 gallons of nitrogen tetroxide. Lockheed Martin and the 3rd Space Launch Squadron evacuated the pad, retreating to the emergency evacuation assembly point. Everyone responded appropriately, and no injuries were reported as a result of the incident. Though the orange cloud resulting from the spill drifted toward the Kennedy Space Center, the local community was never in any danger. Decontamination efforts were completed quickly, and mission essential personnel were allowed back on the pad on 14 August 2003. The pump was replaced, and technicians completed oxidizer loading operations on 17 August 2003.[340] The launch was rescheduled for 21 August 2003, but a suspect fuel tank low-level sensor on Stage II prompted officials to hold the TITAN IVB/CENTAUR at its R-3 day configuration and slip the launch to 6 September 2003. To resolve the sensor problem, the Senior Management Review Team (SMRT) decided to change the TITAN IVB's guidance computer database to "delay looking for low fuel levels until later [in the Stage II] flight." In addition to that action, the wiring connecting the controller to the low-level sensor was rerouted to the outage sensor, and two remaining function sensors were left to control Stage II timing. The software and hardware workarounds were completed successfully, and officials closed out the issue on 2 September 2003.[341]

The countdown finally got underway late on 8 September 2003, and instrumentation problems during the countdown and flight were generally minor. Nevertheless, seven items were more serious in nature, and they deserve special comment. About nine minutes before lift-off, a Cinetheodolite at one of the range's camera sites was Partially Mission Capable (PMC) due to errors in the output azimuth data. After the launch, CSR technicians determined the Truetime Time Code Generator malfunctioned. They replaced it with a spare unit and closed out the anomaly on 10 September 2003. During the plus count, one radar lost "skin" track twice for brief intervals due to sudden drops in signal strength. Analysts blamed tracking angle and flame attenuation for the losses. As an "unavoidable occurrence," they recommended the item be closed out. Three other radars experienced tracking problems while in the beacon mode. Analysts recommended closing out all three items as unavoidable occurrences. Two other anomalies were discovered after the mission. The Cinetheodolite (mentioned earlier) was out of focus during the entire mission. Data quality suffered as a result, and repairs by the Visual Information Technical Contractor (VITC) took a considerable amount of time to complete (e.g., 14 November 2003). Similarly, an Intercept Ground Optical Recorder (IGOR) was out of focus for the duration of the mission. In this instance, the contractor corrected the problem on 30 September 2003 and verified the IGOR was functioning properly on 15 October 2003.[342]

According to a 45 SW Command Post message transmitted to higher headquarters shortly after the event, the TITAN IVB/CENTAUR lifted off the Pad 40 at 0429:00.569Z on 9 September 2003. The 3rd Space Launch Squadron dedicated the launch to Major LeRoy W. Homer (USAFR).[343] The launch was successful, and it provided a happy ending to one of the Cape's longest and most complicated processing operations.[344]

TITAN IVB/IUS (Mission B-39/K-40/IUS-10/DSP-22), 14 February 2004

The purpose of the TITAN IVB-39 mission was to place the Defense Support Program 22 (DSP-22) spacecraft into a predetermined orbit from Complex 40. Like other satellites in the DSP constellation, the spacecraft was designed to detect and report missile launches, space launches, and nuclear detonations. Lockheed Martin chose a TITAN IVB/IUS to carry the DSP-22 into space. According to the flight scenario, the TITAN IVB/IUS vehicle would lift off Pad 40 on an initial flight azimuth of 93 degrees. If all staging sequences were performed correctly, the IUS would separate from the payload six hours, 54 minutes, and 51 seconds after lift-off, and the DSP-22 spacecraft would enter the predetermined orbit.[345]

Technicians rolled the TITAN IVB out to Complex 40 on 22 September 2003, and engineers mated the IUS to the launch vehicle on 24 October 2003. Officials completed the Baseline

Combined Systems Test (BCST) successfully on 14 November 2003. Engineers mated the DSP-22 to the launch vehicle on 9 December 2003. Officials completed the Launch Combined Systems Test on 3 February 2004, and technicians completed fueling operations on the core vehicle (Stage I and Stage II) on 9 February 2004. Oxidizer loading operations were underway on 10 February, and preparations continued for the first countdown on 14 February 2004.[346]

The weather forecast for the upcoming mission indicated there was a 60 percent probability of thick clouds and rain on 14 and 15 February 2004. As events turned out, there was one 33-minute-long unplanned hold during the countdown for weather. The countdown went well, and the launch vehicle lifted off Complex 40 at 1850:00.948Z on 14 February 2004. The launch was successful. The 3 SLS dedicated the DSP-22 mission to Mr. Donald Seib, who started his career at the Cape in 1959. Mr. Seib joined the TITAN launch team in 1968, and he supported missions featuring all variants of the TITAN from TITAN IIIC through TITAN IVB. The DSP-22 flight was his last mission.[347]

TITAN IVB/NUS (Mission B-30/K-36/NROL-16), 30 April 2005

The purpose of the TITAN IVB-30 mission was to place the NROL-16 spacecraft (sponsored by the National Reconnaissance Office) in a predetermined orbit. The launch vehicle carried no upper stage (NUS), and it was launched from Complex 40 on a flight azimuth of 48.5 degrees. Officials initially planned to launch to mission on 20 February 2005, and the TITAN IVB core vehicle was rolled out of the Vertical Integration Building (VIB) on 29 June 2004 with that launch date in mind. Unfortunately, the vehicle's Upgraded Solid Rocket Motors (SRMUs) did not align properly during vehicle operations in the Solid Motor Assembly & Readiness Facility (SMARF) in early July 2004, and workers returned the core vehicle to the VIB for core realignment about a week later.[348]

The rollout to the launch pad (following operations at the SMARF) was scheduled for 20 August 2004, but the actual movement was delayed in anticipation of Hurricane CHARLEY – one of two major hurricanes (the other was IVAN) that struck Florida but failed to cause any serious damage to Patrick or the Cape in 2004. In the meantime, engineers completed the SRMU installation on 24 August 2004. Technicians moved the launch vehicle to Complex 40 on 25 August 2004.[349]

Within a few days, officials began bracing for yet another storm, Hurricane FRANCES. That storm made landfall about 70 miles south of Cape Canaveral over the Labor Day weekend of 4-5 September. A preliminary assessment of FRANCES' impact on the Cape revealed no apparent damage to the TITAN IVB booster on Complex 40 or boosters at other locations. However, the VIB and SMARF had some exterior siding, roof, and door damage as well as sustained water intrusion. Hidden moisture became more apparent in the weeks that followed. On 16 September 2004, a readiness bus umbilical on the TITAN IVB lost isolation due to moisture intrusion from FRANCES. The umbilical pins were dried, and the umbilical was replaced.[350]

Officials began preparing for another hurricane, JEANNE, on 23 September 2004. JEANNE proved to be a virtual copy of FRANCES in both size and power. On 26 September 2004 JEANNE made landfall in approximately the same area of East Central Florida as FRANCES had. Officials were somewhat concerned about a repetition of the moisture problem experienced earlier, but the TITAN IVB's umbilical was none the worse for wear. Technicians accomplished a final retest of the umbilical with satisfactory results on 5 October 2004. Though the schedule for the mission's Launch Combined Systems Test (LCST) slipped from 9 January to 28 January 2005, the launch was still on schedule.[351]

Unfortunately, two more moisture problems emerged on 19 October apparently caused by the hurricanes in September. Technicians discovered six core vehicle ordnance connectors contaminated and corroded, and the "North SRMU" failed to pass isolation checks due to moisture in the SRMU raceway. Technicians replaced the connectors, and all connectors passed operational checks a few days later. The SRMU raceway cables were replaced, and officials completed continuity and isolation tests successfully by 4 November 2004. In the meantime, technicians uncovered two more moisture problems. On 26 October they noticed water intrusion in two subassemblies of the Stage I engine turbo pump assembly fluid heater. Both fluid heaters were inspected for contamination and corrosion, and they were replaced and retested in late November 2004. On 27 October, technicians also found corrosion on some engine instrumentation connectors. Each of the connectors was demated, cleaned, remated, and tested individually. As a precaution, technicians retested the entire Stage I and Stage II engine instrumentation systems following this work. All connectors passed by 17 November 2004, and the issue was closed.[352]

Around the same period, Lockheed Martin requested an indefinite launch date following a "non-nominal" lifting operation in the Solid Motor Assembly Building (SMAB) High Bay on 3 December 2004. On 1 March 2005, a separation circuit failed during a Combined Systems Test (CST). Metallic debris between electric contacts was suspected, and the circuit relay was sent to Denver for analysis. Thorough retesting and a design modification eliminated any lingering concerns.[353]

Officials completed the Launch Combined Systems Test successfully on 13 March, and Lockheed Martin requested a new launch date of 7 April 2005. Officials approved a new launch date of 11 April 2005, but the launch went back to "indefinite" status due to an oxidizer flow pressure problem during a core vehicle (Stage I) fueling operation on 7 April 2005. Troubleshooters suspected a clogged filter, but replacement of the filter did not solve the problem during two attempts on the 7th, and a third attempt on 9 April merely confirmed the source of the problem lay elsewhere. Ultimately, the problem was traced to a damaged shaft and broken bearing in the oxidizer pump. Technicians completed installation of the replacement pump on 20 April, and the Stage I and Stage II oxidizer loading operations were completed successfully on 25 April 2005. In the meantime, officials rescheduled the launch for 30 April 2005, and preparations continued with that launch date in mind.[354]

There were no unplanned holds during the countdown on 29 April, and the TITAN IVB/NUS vehicle lifted off Complex 40 at 0050:00.811Z on 30 April 2005. The launch was successful. Though the TITAN IV program's final TITAN IVB was scheduled for launch from Vandenberg AFB, California, later on, the TITAN IVB/NROL-16 launch constituted the final TITAN IV flight from Cape Canaveral. As such, it signaled the end of an era stretching all the way back to the first TITAN I launch from Complex 15 on 6 February 1959. The spectacular night launch proved a memorable and fitting end to TITAN launch operations at the Cape.[355]

TITAN IV Civil Space Operations

TITAN IVB/CENTAUR, CASSINI Mission, 15 October 1997

Of all the civil space missions launched from Cape Canaveral over the years, only one employed a TITAN IV launch vehicle. In this instance a TITAN IVB/CENTAUR was used to place NASA's CASSINI spacecraft on an interplanetary flight path for its trip to Saturn.[356] The vehicle was launched from Complex 40 on 15 October 1997. The TITAN IVB core vehicle arrived at Cape Canaveral on 13 February 1997. Technicians erected the booster on a transporter on 15 April, and they moved the TITAN IVB to the Solid Motor Assembly & Readiness Facility (SMARF) on 15 May 1997. After workers finished attaching the Upgraded Solid Rocket Motors (SRMUs) to the launch vehicle on 17 May, technicians transported the vehicle from the SMARF to Complex 40 on 30 May 1997. Engineers mated the CENTAUR upper stage to the TITAN IVB on 20 June, and two Terminal Countdown Demonstrations (TCDs) were completed on 5 and 21 August 1997 respectively. Engineers mated the CASSINI spacecraft to the launch vehicle in early September, but it had to be demated on 7 September and remated on 17 September to check out the spacecraft after the Huygens Probe (part of the payload) was over-pressurized.

Technicians installed the payload fairing on 23 September, and officials accomplished the Launch Combined Systems Test (Launch CST) on 2 October 1997. Officials proceeded with final preparations for the launch, which was scheduled for 13 October 1997.[357]

In the meantime, CASSINI's nuclear powerplant had already drawn considerable attention and some heated debate from some private citizens as early as 1996. Though the possibility of an accident during CASSINI's launch was extremely low, protesters considered the launch too risky to proceed. On 26 May 1996, about 125 people from the Florida Coalition for Peace and Justice (FCPJ) staged a peaceful protest near the Cape's west boundary sign on State Route 401. Similar demonstrations involving about 50 people and 95 people respectively were staged in the same location on 14 June and 26 July 1997. Both protests were peaceful, and no one was arrested. The media was represented at both demonstrations, and a "60 Minutes" television crew covered the July 26th protest for national broadcast at a later date. A demonstration at NASA Headquarters drew about 70 protesters on 7 August 1997, and it led to eight arrests. Elsewhere in the world, two women were arrested in London, England in early October for painting anti-CASSINI slogans on the U.S. Embassy. During the same period, several activists were arrested for occupying a U.S. Senator's office. The Hawaii Green Party asked a federal judge in Honolulu, Hawaii to order a halt to the launch via a temporary restraining order. The petition was reviewed in U.S. District Court and denied by Judge David Alan Ezra on 11 October 1997.[358]

On 3 October 1997, NASA Administrator Daniel S. Goldin was given permission to proceed with the CASSINI launch. Cape Security braced for a large-scale demonstration near the west boundary gate on 4 October 1997. Detachments of the Brevard County Sheriff's Office, the Florida Highway Patrol, and the 45th Space Wing's Security Forces Squadron took up positions on the Air Station to manage the situation on the 4th. The protesters began gathering around noon, and the crowd swelled to about 900 before it marched up to the west boundary gate at 3:50 pm. While most of the people remained outside the station, 18 people climbed over the fence only to be arrested by Brevard County deputy sheriffs. In all, 27 people were arrested. The crowd dispersed around 4:30 p.m., and the gate was reopened around 5:05 p.m. Anti-CASSINI pickets continued to show up nearby on 11, 12, 13, and 15 October 1997. Unlike the earlier demonstrations that surrounded the first TRIDENT II launch from Complex 46 in January 1987, the anti-CASSINI demonstrations were not provocative. There were no serious threats of sabotage, and the demonstrations were orderly and properly controlled by their organizers. Nevertheless, law enforcement and military agencies had to be prepared for serious threats on launch day.[359]

The first countdown for the CASSINI launch got underway on 13 October 1997, but officials scrubbed the launch at 0942Z due to high-level winds out of the east. The count was recycled 48 hours, and, this time, circumstances favored the launch. The TITAN IVB/CENTAUR lifted off Pad 40 at 0843:01.111Z on 15 October 1997. The launch was highly successful. One day after the launch, CASSINI was reportedly "cruising along without any problems." NASA was very pleased with the results of the launch, according to Mary Beth Murrill, a spokeswoman for NASA's Jet Propulsion Laboratory.[360]

ENDNOTES

1. Each of the four Castor IVAs was 40 inches in diameter and 36.6 feet long. Each solid rocket motor weighed approximately 25,000 pounds and provided about 98,000 pounds of thrust. Two Castor IVAs were ignited half a second before lift-off, and they continued to burn for about 55 seconds. They separated from the vehicle about 77 seconds after lift-off. The remaining pair of Castor IVAs ignited about 59 seconds into the flight, and they burned for about 55 seconds. The remaining pair of Castor IVAs separated from the vehicle about 117 seconds into the flight.

2. Program Requirements Document, General Dynamics Commercial Launch Services, "FLTSATCOM PRD/OR/RD 3600 (Previously 3200)," 19 Apr 89, pp 1300.1 and 1311.1; Excerpt, General Dynamics Commercial Launch Services, "Mission Planner's Guide for the Atlas Launch Vehicle Family," Rev 1, Mar 89; "A Historical Look at United States Launch Vehicles, 1967 - Present," ANSER's Space Analysis Division (1997), pp II.B-5, II.B-8, II.B-9, II.B-10; 45 SW History, 1 Oct 90 - 31 Dec 91, pp 331, 332; Summary, 45 RANS/DS, "Major Operations FY 93 and First Quarter FY 94," 7 Jun 93 and 31 Jan 94.

3. Excerpt, CSR, "Launch Book, ATLAS I/GOES I, Mission AC-73," Apr 94, Section B; Excerpt, CSR, "Launch Book, ATLAS I/CENTAUR, UHF Follow On 3, Mission AC-76," Jun 94, p B-2; Excerpt, CSR, "Launch Book, ATLAS IIA/CENTAUR, DIRECTV. Mission AC-107," Jul 94, p B-2; Excerpt, CSR, "ATLAS IIAS/CENTAUR, INTELSAT VII, Mission AC-111," Sep 94, p B-3; Excerpt, CSR, "Launch Book, ATLAS IIA/CENTAUR, GOES-M/AC-142 Mission," Rev 4, 20 Jul 01, p B-3; Excerpt, CSR "Launch Book, ATLAS IIAS/CENTAUR, NRO MLV-12/AC-162 Mission," Rev 1, 9 Oct 01, p B-3; Excerpt, CSR, "Launch Book, ATLAS IIAS/CENTAUR, Commercial HISPASAT-1D/AC-159 Mission," Rev 2, 18 Sep 02, pp A-1, A-3, B-3.

4. Excerpt, CSR, "Launch Book, ATLAS I/GOES I, Mission AC-73," Apr 94, Section B; Excerpt, CSR, "Launch Book, ATLAS I/CENTAUR, UHF Follow On 3, Mission AC-76," Jun 94, p B-2; Excerpt, CSR, "Launch Book, ATLAS IIA/CENTAUR, DIRECTV, Mission AC-107," Jul 94, p B-2; Excerpt, CSR, "ATLAS IIAS/CENTAUR, INTELSAT VII, Mission AC-111," Sep 94, p B-3; Excerpt, CSR, "Launch Book, ATLAS IIA/CENTAUR, GOES-M/AC-142 Mission," Rev 4, 20 Jul 01, pp C-3 through C-11, D-1; Excerpt, CSR, "Launch Book, ATLAS IIAS/CENTAUR, NRO MLV-12/AC-162 Mission," Rev 1, 9 Oct 01, pp C-2 through C-5, C-10, D-1; Excerpt, CSR, "Launch Book, ATLAS IIAS/CENTAUR, Commercial HISPASAT-1D/AC-159 Mission," Rev 2, 18 Sep 02, pp A-1, A-3, B-3.

5. When it was completed, the UHF Follow-On constellation was expected to consist of four pairs of operational UHF Follow-On communications satellites and one orbiting spare. One pair of satellites would cover the Continental U.S. and another would cover the Atlantic Ocean. A third pair of satellites would provide communications for the Pacific, and the fourth pair of satellites would cover the Indian Ocean. Each UHF Follow-On spacecraft weighed approximately 6,298 pounds and had an operational service life of 14 years.

6. 45 SW History, 1 Jan – 31 Dec 93, Vol I, pp 197, 198.

7. News Release, 45 SPW/PA, "Commercial Atlas Launch," 23 Mar 93; 45 SW History, 1 Jan – 31 Dec 93, Vol I, pp 198, 200; Summary, 45 RANS/DS, "Major Operations FY-93 and First Quarter FY-94," 7 Jun 93 and 21 Jan 94.

8. As mentioned earlier, the UHF Follow-On constellation was expected to consist of four pairs of operational UHF Follow-On communications satellites and one orbiting spare. Each UHF Follow-On spacecraft weighed approximately 6,298 pounds and had an operational service life of 14 years.

9. 45 SW History, 1 Jan –31 Dec 93, Vol I, p 200; News Release, 45 SPW/PA, "Air Force Atlas II/DSCS III Mission Launched Successfully," 19 Jul 93; "Atlas II launches DSCS III satellite," *45th Space Wing Missileer*, 23 Jul 93, pp 1, 6; Summary, 45 RANS/DS, "Major Operations FY-93 and First Quarter FY-94," 7 Jun 93 and 21 Jan 94.

10. 45 SW History, 1 Jan – 31 Dec 93, Vol I, p 202.

11. 45 SW History, 1 Jan – 31 Dec 93, Vol I, pp 202, 203; Summary, 45 RANS/DS, "Major Operations FY-93 and First Quarter FY-94," 7 Jun 93 and 21 Jan 94.

12. 45 SW History, 1 Jan – 31 Dec 93, Vol I, p 203.

13. 45 SW History, 1 Jan – 31 Dec 93, Vol I, p 203; "Atlas II launch 'picture perfect'," *45th Space Wing Missileer*, 3 Dec 93; News Release, 45 SPW/PA, "Air Force Atlas II/DSCS III Mission Launched Successfully," 28 Nov 93; Summary, 45 RANS/DS, "Major Operations FY-93 and First Quarter FY 94," 7 Jun 93, and 21 Jan 94.

14. 45 SW History, 1 Jan – 31 Dec 94, Vol I, pp 97, 98; "Atlas will head for stars today," *Florida Today* (24 Jun 94).

15. 45 SW History, 1 Jan – 31 Dec 94, Vol I, p 98; "Atlas launch successful," *45th Space Wing Missileer*, 1 Jul 94); Summary, 45 RANS/DS, "Major Operations FY 94," 25 Oct 94, p 3.

16. 45 SW History, 1 Jan – 31 Dec 95, Vol I, pp 95, 96; News Release, Martin Marietta Astronautics, "ATLAS set to launch Navy communication satellite," 26 Jan 95.

17. 45 SW History, 1 Jan – 31 Dec 95, Vol I, p 96; Summary, CSR Range Scheduling, "Major Operations FY-95," 21 Jul 95, p 2.

18. 45 SW History, 1 Jan – 31 Dec 95, Vol I, p 99; News Release, Martin Marietta Astronautics, "ATLAS set to launch Navy communication satellite," 26 Jan 95.

19. 45 SW History, 1 Jan – 31 Dec 95, Vol I, p 99; Summary, CSR Range Scheduling, "Major Operations FY-95," 21 Jul 95, p 3.

20. 45 SW History, 1 Jan – 31 Dec 95, Vol I, p 100; Excerpt, 45 SW/PA, "United States Air Force Mission Information Guide, Atlas/Centaur/DSCS III Mission Number A0116," Jul 95, pp 5, 23, 26, 27; "Atlas team looks to DSCS launch," *45th Space Wing Missileer*, 21 Jul 95.

21. 45 SW History, 1 Jan – 31 Dec 95, Vol I, p 100; "Atlas team looks to DSCS launch," *45th Space Wing Missileer*, 21 Jul 95; Summary, 45 RANS/DOS, "Major Operations FY-95," 27 Nov 95, p 4.

22. 45 SW History, 1 Jan – 31 Dec 95, Vol I, pp 101, 102; News Release, Martin Marietta Astronautics, "ATLAS set to launch Navy communication satellite," 26 Jan 95.

23. 45 SW History, 1 Jan – 31 Dec 95, Vol I, p 102; "Mother Nature causes launch delays for Atlas II, space shuttle Columbia," *45th Space Wing Missileer*, 20 Oct 95, p 1; Summary, 45 RANS/DOUS, "Major Operations FY-96," 23 Feb 96, p 1.

24. 45 SW History, 1 Jan – 31 Dec 96, Vol I, p 84.

25. 45 SW History, 1 Jan – 31 Dec 96, Vol I, pp 84, 85; "Atlas II rocket soars," *45th Space Wing Missileer,* 2 Aug 96, p 1; "Atlas II rocket lofts Navy satellite," *45th Space Wing Missileer*, 2 Aug 96, p 11; Summary, 45 RANS/DOUS, "Major Operations FY-96," 31 Dec 96, p 5.

26. The DSCS III satellite system included the DSCS satellite, an integrated apogee boost system (IABS), a spacecraft adapter and separation system, and all electrical harnesses, connectors, and hardware needed to mate the system to the ATLAS IIA/CENTAUR upper stage. The DSCS III weighed about 1,980 pounds dry, and 2,550 pounds on-orbit. The satellite's body measured 6.8 x 6.3 x 6.4 feet, and its solar array was 38.1 feet across. The DSCS III had an operational life expectancy of 10 years.

27. 45 SW History, 1 Jan – 31 Dec 97, Vol I, p 90.

28. 45 SW History, 1 Jan – 31 Dec 97, Vol I, p 90; Summary, 45 RANS/DOUS, "Major Operations, FY-97," 7 Jan 98, p 1.

29. Mrs. Pearce passed away after 35 years of government service. In October 1993, she was promoted to the position of confidential assistant to the Secretary of the Air Force, where she served secretaries Aldridge, Rice and Widnall. As a token of respect, a decal in honor of Mrs. Pearce was affixed to the payload fairing of the AC-109 vehicle.

30. 45 SW History, 1 Jan – 31 Dec 98, Vol I, pp 81, 82; "Atlas Team to launch its first Nat'l Recon Office payload," *45th Space Wing Missileer,* 23 Jan 98, p 5.

31. 45 SW History, 1 Jan – 31 Dec 98, Vol I, p 82; "Atlas team successfully launches NRO satellite," *45th Space Wing Missileer*, 6 Feb 98, p 1.

32. 45 SW History, 1 Jan – 31 Dec 98, Vol I, pp 84, 85; "Atlas successfully launches satellite," *45th Space Wing Missileer,* 20 Mar 98, p 1.

33. 45 SW History, 1 Jan – 31 Dec 98, Vol I, p 85; Summary, 45 RANS/DOUS, "Major Operations FY-98 and FY-99," 4 Jan 99, p 2.

34. 45 SW History, 1 Jan – 31 Dec 98, Vol I, p 87.

35. 45 SW History, 1 Jan – 31 Dec 98, Vol I, p 87; Summary, 45 RANS/DOUS, "Major Operations FY-98 and FY-99," 4 Jan 99, p 1.

36. In its completed form, the new UHF satellite constellation was expected to consist of eight operational satellites and one on-orbit spare. The spacecraft included EHF, HF, and S-Band communications, and the F-8 through F-10 satellites had a high power, high capacity Global Broadcast Service (GBS) system onboard.

37. 45 SW History, 1 Jan – 31 Dec 99, Vol I, p 71.

38. *Ibid.*

39. The Range was 'Red' for four major problems between 2310Z on the 21st and 0148Z on the 22nd. The first "No-Go" was reported for clouds and rain at 2310Z, and the attendant winds prompted the user to halt tower removal at 0058Z on the 22nd. The second "No-Go" condition involved a 30-inch water deluge line problem that surfaced at 0057Z. The third major problem emerged when the Command Message Encoder Verifier (CMEV) link to Antigua went Non-Mission Capable (NMC) at 0140Z. It remained NMC (prompting the third "No-Go" condition) until the launch was scrubbed. The ATLAS' replacement battery failed its booster checks around 0142Z, and that problem was the last straw. Officials scrubbed the launch at 0148Z.

40. 45 SW History, 1 Jan – 31 Dec 99, Vol I, p 72.

41. 45 SW History, 1 Jan – 31 Dec 2000, Vol I, p 90.

42. 45 SW History, 1 Jan – 31 Dec 2000, Vol I, p 90; "Military launches satellite to join defense network," *Florida Today,* 21 Jan 00, p 4A; "Atlas launched successfully," *45th Space Wing Missileer,* 28 Jan 00, p 4.

43. 45 SW History, 1 Jan – 31 Dec 2000, Vol I, p 97; "Launch shows Atlas team's ingenuity," *45th Space Wing Missileer*, 27 Oct 00, p 1.

44. When the remaining SLEP spacecraft were added to the DSCS constellation, there would be a total of 14 operational and operational spare DSCS III satellites in orbit.

45. 45 SW History, 1 Jan – 31 Dec 2000, Vol I, pp 97, 98; "Atlas launch is tonight," *Florida Today,* 19 Oct 00, p 3A.

46. 45 SW History, 1 Jan – 31 Dec 2000, Vol I, p 98.

47. 45 SW History, 1 Jan – 31 Dec 2000, Vol I, pp 98, 99; "Atlas 2A launches despite problems," *Florida Today,* 20 Oct 00.

48. 45 SW History, 1 Jan – 31 Dec 2000, Vol I, p 99.

49. 45 SW History, 1 Jan – 31 Dec 2000, Vol I, pp 99, 100; "Atlas launch set for Tuesday," *Florida Today,* 4 Dec 00, p 1B.

50. 45 SW History, 1 Jan – 31 Dec 2000, Vol I, p 100; "Atlas mission ends 2000 launch season," *45th Space Wing Missileer,* 8 Dec 00, p 1; "Atlas 2 lifts off with spy satellite," *Florida Today,* 6 Dec 00, p 1A.

51. 45 SW History, 1 Jan – 31 Dec 01, Vol I, pp 91, 92.

52. 45 SW History, 1 Jan – 31 Dec 01, Vol I, p 92; "Wednesday's launch first since upgrades," *45th Space Wing Missileer,* 12 Oct 01, pp 1, 5.

53. 45 SW History, 1 Jan – 31 Dec 04, Vol I, p 74; "Into History," *Aviation Week & Space Technology,* 13 Sep 04, p 64; "Era ends for launch vehicle, pad," *45th Space Wing Missileer,* 27 Aug 04, p 1.

54. 45 SW History, 1 Jan – 31 Dec 04, Vol I, p 74.

55. 45 SW History, 1 Jan – 31 Dec 04, Vol I, pp 74, 75.

56. 45 SW History, 1 Jan – 31 Dec 04, Vol I, p 75; "A secret National Reconnaissance Office spacecraft…," *Aviation Week & Space Technology,* 6 Sep 04, p 18.

57. The TELSTAR 4 spacecraft was built by Martin Marietta Astro Space Division for AT&T to provide telecommunications services for all 50 U.S. states, Puerto Rico, and the Virgin Islands. The spacecraft weighed 7,250 pounds, and it was equipped with eight solar array panels. The TELSTAR 4 featured Ku-Band and C-Band feed assemblies, earth sensors, and a single omni antenna. The spacecraft was equipped with Liquid Apogee Engines (LAEs) designed to burn hydrazine with nitrogen tetroxide as the oxidizer. The payload fairing used for the TELSTAR 4 was the large 14-foot-diameter ATLAS/CENTAUR fairing.

58. 45 SW History, 1 Jan – 31 Dec 93, Vol I, p 204; "Atlas IIAS lifts Telstar into orbit," *45th Space Wing Missileer*, 17 Dec 93.

59. 45 SW History, 1 Jan – 31 Dec 93, Vol I, pp 204, 206; News Release, 45 SPW/PA, "Commercial Atlas Launched," 15 Dec 93.

60. The first DBS satellite was launched on an ARIANE 4 from Kourou, French Guiana, on 17 December 1993. Both DBS satellites were HS 601s build by GM Hughes Electronics. Each satellite featured sixteen 120-watt Ku-band transponders, which would be used to relay laserdisc quality video and compact disc sound. Each satellite weighed about 3,800 pounds. The spacecraft measured 23.3 feet across its transmit antennas and 86 feet across its deployed solar arrays. DIRECTV's operational life expectancy was 12 years. Programming was directed from a digital broadcast facility at the DIRECTV Broadcast Center in Castle Rock, Colorado.

61. 45 SW History, 1 Jan – 31 Dec 94, Vol I, pp 98, 100; "Delivering DirectTV," *Uplink* (Summer 1992), p 9; Fact Sheet, Hughes Space and Communications Company, "Satellites to Deliver TV Direct to Home Viewers," undated; Summary, 45 RANS/DS, "Major Operations FY 94,"25 Oct 94, p 3.

62. The ATLAS IIAS' Thiokol Castor IVA solid rocket boosters increased the launch vehicle's total thrust to 620,500 pounds. This power was sufficient to boost the payload into geosynchronous transfer orbit. Space Systems/Loral built the INTELSAT VII under a contract with the International Telecommunications Satellite (INTELSAT) Organization. The satellite provided voice, video and data transmission services, and it had an operational life expectancy of 10.9 years. The INTELSAT 703 was the second spacecraft in the INTELSAT VII series, and it brought INTELSAT's global network to 22 orbiting satellites. The INTELSAT constellation served more than 200 countries and territories around the world.

63. 45 SW History, 1 Jan – 31 Dec 94, Vol I, p 100, 101; News Release, Martin Marietta Astronautics, "ATLAS IIAS Successfully Launches INTELSAT 703 Communications Satellite," ca. 7 Oct 94; "Atlas sends Intelsat into orbit," *Florida Today*, 7 Oct 94); "Atlas rocket takes flight," *Florida Today*, 6 Oct 94; Summary, 45 RANS/DS, "Major Operations FY 95,"3 Jan 95, p 1.

64. 45 SW History, 1 Jan – 31 Dec 94, Vol I, pp 101, 102; "ORION Satellite Corp," *Janes's Spaceflight Directory*, CY 93-94, p 381; "Atlas Boosts ORION Satellite," *Aviation Week & Space Technology*, 5 Dec 94, p 26.

65. 45 SW History, 1 Jan – 31 Dec 94, Vol I, p 102; "Tuesday's Atlas launch on track," *Florida Today*, 21 Nov 94; "Atlas rocket launch set for Tuesday," *Florida Today*, 24 Nov 94; "Atlas rocket's second launch try Tuesday," *Florida Today,* 28 Nov 94; "Atlas cuts eerie swath across sky," *Florida Today*, 30 Nov 94; Summary, 45 RANS/DS, "Major Operations FY 95," p 1.

66. 45 SW History, 1 Jan – 31 Dec 95, Vol I, p 95.

67. 45 SW History, 1 Jan – 31 Dec 95, Vol I, p 95; "Atlas launch starts year off right for wing," *45th Space Wing Missileer*, 13 Jan 95, p 1; Summary, CSR Range Scheduling, "Major Operations FY-95," 21 Jul 95, p 2.

68. 45 SW History, 1 Jan – 31 Dec 95, Vol I, p 96.

69. 45 SW History, 1 Jan – 31 Dec 95, Vol I, pp 96, 97; News Release, 45 SW/PA, "Successful Commercial ATLAS II Launches from Cape Canaveral," 22 Mar 95; Summary, CSR Range Scheduling, "Major Operations FY-95," 21 Jul 95, p 2.

70. 45 SW History, 1 Jan – 31 Dec 95, Vol I, p 97.

71. 45 SW History, 1 Jan – 31 Dec 95, Vol I, pp 97, 98; News Release, 45 SW/PA, "Successful Commercial ATLAS IIA Launches from Cape Canaveral," 7 Apr 95; Summary, CSR Range Scheduling, "Major Operations FY-95," 21 Jul 95, p 2.

72. 45 SW History, 1 Jan – 31 Dec 95, Vol I, p 101.

73. 45 SW History, 1 Jan – 31 Dec 95, Vol I, p 101; Summary, 45 RANS/DOS, "Major Operations FY-95," 27 Nov 95, p 4.

74. 45 SW History, 1 Jan – 31 Dec 95, Vol I, pp 103, 104; "Atlas launches Galaxy III satellite, Delta XTE put on hold until Dec. 27," *45th Space Wing Missileer*, 22 Dec 95, p 1.

75. 45 SW History, 1 Jan – 31 Dec 95, Vol I, p 104; "Atlas launches Galaxy III satellite, Delta XTE put on hold until Dec.27," *45th Space Wing Missileer*, 22 Dec 95, p1; Summary, 45 RANS/DOUS, "Major Operations, FY 96, 23 Feb 96, p1.

76. 45 SW History, 1 Jan – 31 Dec 96, Vol I, pp 81, 82; "Atlas rocket lights up night skies," *45th Space Wing Missileer,* 9 Feb 96, p 1.

77. 45 SW History, 1 Jan – 31 Dec 96, Vol I, p 82; "Atlas rocket lights up night skies," *45th Space Wing Missileer,* 9 Feb 96, p 1; Summary, 45 RANS/DOUS, "Major Operations FY-96," 9 Jul 96, p 2.

78. 45 SW History, 1 Jan – 31 Dec 96, Vol I, pp 82. 83.

79. 45 SW History, 1 Jan – 31 Dec 96, Vol I, p 83; Summary, 45 RANS/DOUS, "Major Operations FY-96," 9 Jul 96, p 3.

80. 45 SW History, 1 Jan – 31 Dec 96, Vol I, pp 83, 84; Excerpt, CSR, "Launch Book, ATLAS I/CENTAUR, SAX, AC-078 Mission," Apr 96, p A-1.

81. 45 SW History, 1 Jan – 31 Dec 96, Vol I, p 84; Summary, 45 RANS/DOUS, "Major Operations FY-96," 9 Jul 96, p 4.

82. 45 SW History, 1 Jan – 31 Dec 96, Vol I, p 85; "Delayed Atlas begins three-launch swing for Space Wing," *45th Space Wing Missileer*, 30 Aug 96.

83. 45 SW History, 1 Jan – 31 Dec 96, Vol I, pp 85, 86; Summary, 45 RANS/DOUS, "Major Operations FY-96," 31 Dec 96, p 5.

84. 45 SW History, 1 Jan – 31 Dec 96, Vol I, p 86;"Atlas set to boost European satellite," *45th Space Wing Missileer*, 8 Nov 96, p 11.

85. 45 SW History, 1 Jan – 31 Dec 96, Vol I, pp 86, 88; Summary, 45 RANS/DOUS, "Major Operations FY-97," 31 Dec 96, p 1.

86. The first INMARSAT-3 payload was launched on the AC-122 mission from Pad 36A on 3 April 1996.

87. 45 SW History, 1 Jan – 31 Dec 96, Vol I, p 88.

88. 45 SW History, 1 Jan – 31 Dec 96, Vol I, p 88; Summary, 45 RANS/DOUS, "Major Operations FY-97," 31 Dec 96, p 1.

89. 45 SW History, 1 Jan – 31 Dec 97, Vol I, pp 84, 85.

90. 45 SW History, 1 Jan – 31 Dec 97, Vol I, p 85; Summary, 45 RANS/DOUS, "Major Operations, FY-97," 2 Oct 97, p 2.

91. 45 SW History, 1 Jan – 31 Dec 97, Vol I, p 85;"Another Atlas Success!" *45th Space Wing Missileer,* 14 Mar 97.

92. 45 SW History, 1 Jan – 31 Dec 97, Vol I, pp 85, 87; "Another Atlas Success!," *45th Space Wing Missileer,* 14 Mar 97; Summary, 45 RANS/DOUS, "Major Operations, FY-97," 2 Oct 97, p 2

93. 45 SW History, 1 Jan – 31 Dec 97, Vol I, p 88; "Atlas, Delta complete successful launches," *45th Space Wing Missileer,* 1 Aug 97, p 1.

94. 45 SW History, 1 Jan – 31 Dec 97, Vol I, p 88; "Atlas, Delta complete successful launches," *45th Space Wing Missileer,* 1 Aug 97, p 1; Summary, 45 RANS/DOUS, "Major Operations, FY-97," 2 Oct 97, p 4.

95. 45 SW History, 1 Jan – 31 Dec 97, Vol I, p 88.

96. 45 SW History, 1 Jan – 31 Dec 97, Vol I, p 89; Summary, 45 RANS/DOUS, "Major Operations, FY-97," 2 Oct 97, p 5.

97. 45 SW History Office, 1 Jan – 31 Dec 97, Vol I, p 89.

98. 45 SW History Office, 1 Jan – 31 Dec 97, Vol I, p 89; Summary, 45 RANS/DOUS, "Major Operations, FY-97," 7 Jan 98, p 1.

99. 45 SW History, 1 Jan – 31 Dec 97, Vol I, pp 90, 91.

100. 45 SW History, 1 Jan – 31 Dec 97, Vol I, p 91; Summary, 45 RANS/DOUS, "Major Operations, FY-97," 7 Jan 98, p 1.

101. 45 SW History, 1 Jan – 31 Dec 98, Vol I, pp 82, 84; "Atlas successfully launches satellite," *45th Space Wing Missileer,* 6 Mar 98, p 1.

102. 45 SW History, 1 Jan – 31 Dec 98, Vol I, p 84; Summary, 45 RANS/DOUS, "Major Operations FY-98 and FY-99," 4 Jan 99, p 2.

103. 45 SW History, 1 Jan – 31 Dec 98, Vol I, p 85; "Atlas successfully launches satellite," *45th Space Wing Missileer,* 26 Jun 98, p 1.

104. 45 SW History, 1 Jan – 31 Dec 98, Vol I, pp 85, 86.

105. 45 SW History, 1 Jan – 31 Dec 98, Vol I, p 86; "Atlas extends its successful launch streak," *45th Space Wing Missileer*, 16 Oct 98, p 1.

106. 45 SW History, 1 Jan – 31 Dec 98, Vol I, p 86.

107. 45 SW History, 1 Jan – 31 Dec 99, Vol I, p 67.

108. *Ibid.*

109. 45 SW History, 1 Jan – 31 Dec 99, Vol I, p 69.

110. *Ibid.*

111. 45 SW History, 1 Jan – 31 Dec 99, Vol I, pp 69, 70; "3rd Space Launch Squadron back on track with Atlas launch," *45th Space Wing Missileer,* 1 Oct 99, p 1; "Atlas Returns With EchoStar," *Aviation Week & Space Technology,* 27 Sep 99, pp 38.

112. Lockheed Martin grounded the ATLAS II fleet following the DELTA III/ORION-3 mishap on 5 May 1999. That mishap was caused by a burst CENTAUR combustion chamber. Since the CENTAUR upper stages on the DELTA III and ATLAS IIAS were similar, Lockheed Martin took no chances until an in-depth investigation cleared the ATLAS IIAS/CENTAUR upper stages from similar mishaps. The ban on ATLAS IIAS/CENTAUR missions was lifted by Lockheed Martin around 31 August 1999.

113. Two small boats were detected inside the one-boat contour at 0426Z. A Coast Guard cutter was dispatched to clear them out of the area. As the two boats left, a ship was spotted some distance away crossing into the no-ship contour line. The cutter cleared that ship from the line, but another ship crossed the line. The second ship continued to "foul" the range until 0543Z. The countdown resumed at 0557Z.

114. 45 SW History, 1 Jan – 31 Dec 99, Vol I, p 70; "3rd Space Launch Squadron back on track with Atlas launch," *45th Space Wing Missileer,* 1 Oct 99, p 1; "Atlas Returns With EchoStar," *Aviation Week & Space Technology,* 27 Sep 99, pp 38; "Lockheed Martin's Centaur RL10 engines cleared for flight," *Florida Today* Space Online, 31 Aug 99; "Atlas rockets get green light for next launches," *Florida Today* Space Online, 31 Aug 99; "Lockheed Martin scrubs Atlas launch," *Florida Today* Space Online, 12 Sep 99; "Atlas nearing launch resumption; Lockheed Martin encounters more engine troubles," *Florida Today* Space Online, 12 Sep 99.

115. 45 SW History, 1 Jan – 31 Dec 2000, Vol I, pp 90, 91.

116. The bridge crane on Pad 36B broke down on 24 January, and it delayed the spacecraft mating and the Composite Electrical Readiness Test (CERT) by one day each. Technicians repaired the crane and brought it back in service on 25 January, but excessive winds on the 26th delayed the spacecraft mate until 27 January 2000. Engineers mated the spacecraft to the launch vehicle on 27 January, and officials completed the CERT on 28 January 2000.

117. 45 SW History, 1 Jan – 31 Dec 2000, Vol I, p 91; "Atlas shakes Space Coast again," *45th Space Wing Missileer,* 11 Feb 00, p 4.

118. 45 SW History, 1 Jan – 31 Dec 2000, Vol I, p 96.

119. The Torus crane was used to lift the SRMs so they could be attached to the ATLAS II vehicle. The crane's electrical controls were out of service from 14 through 21 June 2000. Lockheed Martin replaced the broken parts and checked the system out on 21 June, and the SRM mate was completed on 22 June 2000.

120. 45 SW History, 1 Jan – 31 Dec 2000, Vol I, pp 96, 97.

121. Boeing Satellite Systems, Inc. built the ICO-A1 (F2) spacecraft for the London-based New ICO (formerly ICO Global Communications). The satellite was a version of the popular HS-601 series communications spacecraft. It weighed 6,050 pounds, and it had an operational lifespan of 12 years. Sea Launch (a launch service company partly owned by Boeing) launched the first ICO-A1 spacecraft unsuccessfully on an earlier mission. The AC-156 mission marked the actual beginning of the ICO-A1 network. As envisioned, the ICO-A1 satellite constellation would consist of 10 operational spacecraft and two spares orbiting in two orthogonal planes. Each satellite would circle Earth every six hours, and the orbits would overlap sufficiently to ensure at least two satellites were in view of a customer and an access node at all times. New ICO's constellation was designed to provide third-generation wireless services to a wide variety of customers (e.g., small/medium businesses, oil/gas and maritime industries, government agencies, and individual consumers).

122. 45 SW History, 1 Jan – 31 Dec 01, Vol I, p 88.

123 45 SW History, 1 Jan – 31 Dec 01, Vol I, pp 88, 89; "Atlas Launches ICO, But 'Will It Fly?'," *Aviation Week & Space Technology*, 25 June 01, p 55; "Atlas rocket lofts communications satellite," *45th Space Wing Missileer*, 22 Jun 01, p 6.

124. 45 SW History, 1 Jan – 31 Dec 02, Vol I, p 86; "World News Roundup," *Aviation Week & Space Technology*, 23 Sep 02, p 18.

125. 45 SW History, 1 Jan – 31 Dec 02, Vol I, p 88; "World News Roundup," *Aviation Week & Space Technology*, 23 Sep 02, p 18.

126. 45 SW History, 1 Jan – 31 Dec 04, Vol I, p 69.

127. *Ibid.*

128. 45 SW History, 1 Jan – 31 Dec 04, Vol I, p 70.

129. 45 SW History, 1 Jan – 31 Dec 04, Vol I, pp 70, 71.

130. 45 SW History, 1 Jan – 31 Dec 04, Vol I, p 71.

131. *Ibid.*

132. 45 SW History, 1 Jan – 31 Dec 04, Vol I, p 72.

133. *Ibid.*

134. 45 SW History, 1 Jan – 31 Dec 04, Vol I, p 72.

135. The GOES I through M series carried an Imager and a Sounder as its two major instruments. Those items were developed under a subcontract with the Aerospace/Communications Division of the ITT Corporation. The satellite used the instruments to acquire high-resolution visible and infrared data, temperature, and moisture profiles of Earth's atmosphere. The GOES I satellite pictured about one-fourth of the globe every thirty minutes, and it transmitted its data to ground stations for processing and rebroadcast to primary weather services in the U.S. and around the world. The pictures and data assisted scientists in weather detection and prediction, observation of ocean currents, river water levels and other meteorological evaluations. The satellite also relayed distress signals from people, aircraft, and vessels to search and rescue ground stations in the Search and Rescue Satellite Aided Tracking (SARSAT) system. The GOES I spacecraft weighed 2,161 pounds dry, and it carried approximately 2,479 pounds of fuel. The main body of the satellite was 6.5 x 7 x 7.6 feet. With its solar array and solar sail deployed, the GOES I was 88.3 feet long. The satellite had an operational life expectancy of five years.

136. 45 SW History, 1 Jan – 31 Dec 94, Vol I, pp 96, 97; Excerpt, General Dynamics Space Systems Division, "GDSS-TP-GO-89-002, GOES Summary," undated, pp 1-1, 3-1; Excerpt, NOAA, "Mission Overview, Geostationary Operational Environmental Satellite," undated, pp 2, 3, 5.

137. 45 SW History, 1 Jan – 31 Dec 94, Vol I, pp 97; Summary, 45 RANS/DS, "Major Operations FY-94,"8 Jun 94, p 2; "Atlas GOES a success," *45th Space Wing Missileer*, 22 Apr 94, p 1.

138. 45 SW History, 1 Jan – 31 Dec 95, Vol I, p 98.

139. 45 SW History, 1 Jan – 31 Dec 95, Vol I, pp 98, 99; New Release, 45 SW/PA, "Successful ATLAS I Launch from Cape Canaveral," 23 May 95; Summary, CSR Range Scheduling, "Major Operations FY-95," 21 Jul 95, p 3.

140. 45 SW History, 1 Jan – 31 Dec 95, Vol I, pp 102, 103; Excerpt, CSR, "Launch Book, ATLAS IIAS/CENTAUR, SOHO, AC-121 Mission," Nov 95, p A-1.

141. 45 SW History, 1 Jan – 31 Dec 95, Vol I, p 103; Summary, 45 RANS/DOUS, "Major Operations FY-96," 23 Feb 96, p 1.

142. 45 SW History, 1 Jan – 31 Dec 97, Vol I, p 87.

143. 45 SW History, 1 Jan – 31 Dec 97, Vol I, p 87; Summary, 45 RANS/DOUS, "Major Operations, FY-97," 2 Oct 97, p 3.

144. The spacecraft featured a five-spectral channel imager, a 19-spectral channel atmospheric sounder, and a space environment monitoring system to measure magnetic field strength, solar x-ray flux, and charged particle populations.

145. 45 SW History, 1 Jan – 31 Dec 2000, Vol I, pp 91, 92.

146. Lockheed Martin grounded the ATLAS II fleet following the DELTA III/ORION-3 mishap on 5 May 1999. That mishap was caused by a burst CENTAUR combustion chamber. Since the CENTAUR upper stages on the DELTA III and ATLAS IIAS were similar, Lockheed Martin took no chances until an in-depth investigation cleared the ATLAS IIAS/CENTAUR upper stages from similar mishaps. The ban on ATLAS IIAS/CENTAUR missions was lifted by Lockheed Martin around 31 August 1999.

147. 45 SW History, 1 Jan – 31 Dec 2000, Vol I, p 92.

148. 45 SW History, 1 Jan – 31 Dec 2000, Vol I, p 93.

149. 45 SW History, 1 Jan – 31 Dec 2000, Vol I, pp 93, 94; "Advanced TDRS To Enhance Civil, Military Space Relay," *Aviation Week & Space Technology,* 3 Jul 00, p 39.

150. Pratt & Whitney had just completed an extended investigation of a thrust control valve anomaly involving an RL-10B-2 engine in production for the DELTA III program. Since the CENTAUR stage of the ATLAS IIA/CENTAUR carried two RL10A-4s (a slightly different version of the basic RL-10 engine), NASA took a serious interest in the manufacturer's findings.

151. 45 SW History, 1 Jan – 31 Dec 2000, Vol I, p 94; "Atlas launch puts NASA satellite into orbit," *45th Space Wing Missileer,* 7 Jul 00; Memo, 45 RANS/DOOA, "Atlas IIA/TDRS-H Post Launch Report," 6 Jul 00.

152. According to the flight scenario, the CENTAUR upper stage would place the GOES-M in a super-synchronous transfer orbit. That orbit was approximately 3,000 miles higher than the geosynchronous parking altitude the spacecraft would occupy when it went into operation. The higher altitude saved fuel aboard the spacecraft during final positioning on orbit.

153. 45 SW History, 1 Jan – 31 Dec 01, Vol I, p 89; "New Goes Metsat Orbited," *Aviation Week & Space Technology*, 30 Jul 01, p 32.

154. The range went 'Red' for a violation of the thick cloud rule at 0142Z, but it was the lightning strike on Pad 36B that actually precipitated the scrub. Following the strike, the GOES-M temperature sensor on Pad 36A started "showing glitches," and the user requested the scrub and a 24-hour turnaround for the next launch attempt.

155. 45 SW History, 1 Jan – 31 Dec 01, Vol I, p 91.

156. 45 SW History, 1 Jan – 31 Dec 02, Vol I, pp 83, 84; Advanced TDRS To Enhance Civil, Military Space Relay," *Aviation Week & Space Technology*, 3 Jul 00, p 39; "TDRS-I Satellite Prepped for Launch," *Space News*, 25 Feb 02, p 16.

157. 45 SW History, 1 Jan – 31 Dec 02, Vol I, p 84.

158. The TDRS-H spacecraft was on-orbit. It had been launched successfully from Pad 36A on 30 June 2000.

159. 45 SW History, 1 Jan – 31 Dec 02, Vol I, p 84.

160. 45 SW History, 1 Jan – 31 Dec 02, Vol I, pp 85, 86.

161. 45 SW History, 1 Jan – 31 Dec 02, Vol I, p 86.

162. 45 SW History 1 Jan – 31 Dec 02, Vol I, p 88; "Advanced TDRS To Enhance Civil, Military Space Relay," *Aviation Week & Space Technology*, 3 Jul 00, p 39.

163. 45 SW History 1 Jan – 31 Dec 02, Vol I, pp 88, 89.

164. 45 SW History 1 Jan – 31 Dec 02, Vol I, p 89.

165. 45 SW History, 1 Jan – 31 Dec 2000, Vol I, pp 100, 102.

166. The ATLAS IIIA/CENTAUR was equipped with only one of the new engines, but the ATLAS IIIB/CENTAUR could be equipped with either one or two RL10A-4-1 engines. Each RL10A-4-1 produced 22,300 pounds of thrust. The ATLAS IIIA/CENTAUR upper stage measured 10 x 33 feet. The ATLAS IIIB/CENTAUR upper stage had the same diameter as the ATLAS IIIA/CENTAUR upper stage, but it was five and one-half feet longer (e.g.,10 x 38.5 feet).

167. 45 SW History, 1 Jan - 31 Dec 2000, pp 100, 101, 103; Summary, 45 SW/HO, "Eastern Range Launches, 1 Jan 03 - 1 Jan 04," printed 18 Feb 04; Summary, Lockheed Martin, "Atlas - Facts," 27 Jul 00, p 2; Summary, Lockheed Martin Space Systems Company, "Atlas Profile," undated, pp 1, 2, 3; Summary, Lockheed Martin, "Atlas," 4 Sep 02.

168. Summary, Lockheed Martin Space Systems Company, "Atlas Profile," undated, pp 1, 2, 3; Excerpt, CSR, "Launch Book, ATLAS IIIB/CENTAUR, Commercial ASIASAT-4/AC-205 Mission," Rev 2, 9 Apr 03, p B-3; Excerpt, CSR, "Launch Book, ATLAS IIIB/CENTAUR, Commercial UHF F-11/AC-203 Mission," 2 Dec 03, p B-3.

169. Excerpt, CSR, "Launch Book, ATLAS IIIA/CENTAUR, Commercial EUTELSAT W4/AC-201 Mission," Rev 2, 11 May 00, pp C-3 through C-6, C-11, D-1; Excerpt, CSR, "Launch Book, ATLAS IIIB/CENTAUR, Commercial ASIASAT-4/AC-205 Mission," Rev 2, 9 Apr 03, pp C-2 through C-5, C-10, C-12, D-1; Excerpt, CSR, "Launch Book, ATLAS IIIB/CENTAUR, Commercial UHF F-11/AC-203 Mission," 2 Dec 03, pp C-2 through C-6, C-9, C-11, D-1.

170. 45 SW History, 1 Jan – 31 Dec 06, Vol I, pp 30, 31.

171. 45 SW History, 1 Jan -31 Dec 06, Vol I, p 31; "Historic Complex 36 towers, toppled," *45th Space Wing Missileer*, 22 Jun 07, p 3.

172. Lockheed Martin was proud of the new vehicle, and the corporation wished to honor its arrival with a ceremony marking the beginning of ATLAS III operations at Complex 36. Military officials, contractors and civilian guests attended the event.

173. "New Atlas III booster arrives at CCAS," *45th Space Wing Missileer,* 12 Mar 99, p 4; "Atlas III Flight Tightens LockMart, Boeing Faceoff," *Aviation Week & Space Technology,* 29 May 00; 45 SW History, 1 Jan – 31 Dec 99, p 91.

174. 45 SW History, 1 Jan – 31 Dec 2000, Vol I, pp 102, 103; "Atlas III Flight Tightens LockMart, Boeing Faceoff," *Aviation Week & Space Technology,* 29 May 00.

175. 45 SW History, 1 Jan – 31 Dec 02, Vol I, p 91; "First Atlas 3B Launches EchoStar 7 Satellite," *SpaceNews,* 25 Feb 02, p 12.

176. Technicians erected the ATLAS IIIB booster on Pad 36B as planned on 29 October 2001, but a vibration test failure at the factory in late November 2001 delayed the EchoStar VII's arrival at the Cape until 13 January 2002.

177. 45 SW History, 1 Jan – 31 Dec 02, Vol I, pp 91, 92; "First Atlas 3B Launches EchoStar 7 Satellite," *SpaceNews,* 25 Feb 02, p 12.

178. The spacecraft's C-Band transponders would service an area ranging from New Zealand to portions of the Middle East. The Ku-Band transponders would focus on China, North and South Korea, Taiwan, Hong Kong, Australia, and New Zealand.

179. 45 SW History, 1 Jan – 31 Dec 03, Vol I, p 78; "AsiaSat Heads for GEO," *Aviation Week & Space Technology*, 21 Apr 03, p 54.

180. 45 SW History, 1 Jan – 31 Dec 03, Vol I, p 79.

181. 45 SW History, 1 Jan – 31 Dec 2004, Vol I, pp 78, 79; "Mobile Milestone," *Aviation Week & Space Technology*, 22 Mar 04, p 35.

182. 45 SW History, 1 Jan – 31 Dec 2004, Vol I, p 79; "Nightime lights," *45th Space Wing Missileer*, 19 Mar 04, p 1.

183. The UHF Follow-On satellites replaced Navy FLTSATCOM satellites that began providing global military communications in the late 1970s. Beginning on 25 March 1993, the UHF F-1 through UHF F-10 spacecraft were launched from Cape Canaveral on a series of ATLAS I/CENTAUR, ATLAS II/CENTAUR, and ATLAS IIA/CENTAUR flights. The UHF F-11 spacecraft was the first satellite in the UHF Follow-On series to be launched aboard an ATLAS IIIB/CENTAUR space vehicle.

184. 45 SW History, 1 Jan – 31 Dec 03, Vol I, pp 79, 81.

185. 45 SW History, 1 Jan – 31 Dec 03, Vol I, p 81.

186. 45 SW History, 1 Jan – 31 Dec 03, Vol I, p 82.

187. 45 SW History, 1 Jan – 31 Dec 03, Vol I, pp 82, 83.

188. 45 SW History, 1 Jan – 31 Dec 2005, Vol I, p 71; Article (U), "Final Atlas 3 launches," *Spacetoday.net,* 3 Feb 05; Article (U), "ILS Launches NRO Payload," *SatNews Daily,* 3 Feb 05.

189. 45 SW History, 1 Jan – 31 Dec 2005, Vol I, p 71.

190. 45 SW History, 1 Jan – 31 Dec 2005, Vol I, pp 71, 72.

191. 45 SW History, 1 Jan – 31 Dec 2005, Vol I, p 72; Article (U), "Atlas launch marks end of era," *45th Space Wing Missileer*, p 4, 11 Feb 05.

192. Unlike earlier efforts to develop a new generation of space launch vehicles from a "clean sheet of paper," the EELV concept capitalized on proved hardware found in the current generation of DELTA II, ATLAS II/CENTAUR, and TITAN IV vehicles. The goal was the creation of a cost-efficient and reliable family of right-sized spacelifters based on standardized fairings, liquid core boosters, upper stages, and solid rockets. Standard payload interfaces were touted as another way to save money and improve efficiency, though success in that area was yet to be proven.

193. Briefing Slides, Brigadier General DeKok, AFSPC/XPX, "Booster Roadmap," 15 Dec 94; Memo, Mr. Edmond F. Gormel, 45 SW/XP, to 45 SW Offices, "Evolved Expendable Launch Vehicle (EELV) Procurement," 16 Mar 95; "The evolution of spacelift," *The Guardian*, Jan 96, pp 6, 7; Briefing Slides, SMC/MV, "EELV Government Day," 18 Oct 95.

194. "Lockheed Martin and McDonnell Douglas move forward in EELV competition," *45th Space Wing Missileer*, 10 Jan 97, p 11; Discussion, M. Cleary, 45 SW/HO, with Mr. Frank Mann, CSR, 6 Nov 97; "Countdown to the Future," *The Guardian*, Aug 96; Discussion, M. Cleary, 45 SW/HO, with Lt. Colonel Wayne Eleazer, 45 SW/XPE, 9 Jan 98.

195. 45 SW History, 1 Jan - 31 Dec 97, Vol I, p 48; "Pentagon OKs Two EELVs," *Aviation Week & Space Technology*, 10 Nov 97, p 41; "DoD Reverses Strategy on Evolved Expendable Launcher," *Launchspace*, Apr 98, pp 4, 5; News Release, Defenselink, "Evolved Expendable Launch Vehicle Contract Award," 16 Oct 98; "EELV Win Boosts Boeing Launch Plans," *Aviation Week & Space Technology*, 26 Oct 98, p 71.

196. Designers at Pratt-Whitney and NPO Energomash expected the RD-180 engine to provide 860,200 pounds of thrust at sea level, and the new engine could be throttled from 40 percent to

100 percent of rated thrust. Pratt Whitney intended to build the RD-180 for government missions, but NPO Energomash provided the RD-180s used on all the early ATLAS V launches.

197. "DoD Reverses Strategy on Evolved Expendable Launcher," *Launchspace*, Apr 98, pp 4, 5; "Lockheed Martin EELV Designed for Flexibility," *Aviation Week & Space Technology*, 20 Jul 98, p 57.

198. "Lockheed Martin EELV Designed for Flexibility," *Aviation Week & Space Technology*, 20 Jul 98, pp 56, 57.

199. Detachment 8 also became responsible for integrating the removal of other equipment owned by the Wing, CSR, Bionetics, Johnson Controls, Sverdrup, EG&G and Boeing into the master schedule for equipment removal on Complex 41. By the summer of 1998, Detachment 8 was sponsoring biweekly meetings to coordinate deactivation and equipment issues for Complex 41.

200. 45 SW History, 1 Jan – 31 Dec 98, Vol I, pp 47, 48.

201. The situation begs a question: was a permit required? According to the Webb-McNamara Agreement of 17 January 1963 and the Huston-Debus Memorandum of Agreement (signed 19 and 23 October 1964), all property within the security fence around Complex 41, supporting facilities, utilities, access roads, parking lots, paved areas, the rail transport system, and an area 200 feet wide on each side of the tracks extending from the Cape boundary to the fenced complex *were not included in the Merritt Island Launch Area*. In effect, the 1963 and 1964 agreements extended the Cape boundary into Merritt Island to surround Complex 41 as part of the Air Force's territory.

202. 45 SW History, 1 Jan – 31 Dec 98, Vol I, p 48.

203. 45 SW History, 1 Jan – 31 Dec 98, Vol I, p 49.

204. 45 SW History, 1 Jan – 31 Dec 98, Vol I, pp 49, 50.

205. 45 SW History, 1 Jan – 31 Dec 98, Vol I, p 50.

206. E-Mail, Suzanne Schulman, 45 CES/CEL, to 45SWHO, "Input for Cape History Books," 11 Jan 2000.

207. 45 SW History, 1 Jan -31 Dec 99, Vol I, pp 20, 21;"3rd Space Launch Squadron clears way for space progress," *45th Space Wing Missileer,* 15 Oct 99, pp 1, 6; "Launch complex becomes scrap in seconds," *Florida Today,* 15 Oct 99, pp B-1, B-2; "Launch towers at SLC 41 toppled," *45th Space Wing Missileer,* 22 Oct 99, pp 8, 9.

208. The five-foot thick slab rested on 65-foot-deep pilings to ensure a solid foundation for the ATLAS V launch vehicle. In compensation for the 10.4 acres of "low quality" wetlands required in the VIF construction area, Lockheed Martin agreed to pay the U.S. Fish and Wildlife Service to restore/create more than 54 acres of high-quality wetlands along the Banana River south of the VIF site. Wetland restoration was underway in 1999, and it was completed in 2000.

209. As its name implied, the VIF was used to stack and integrate the ATLAS V on the Mobile Launch Platform. In accordance with Lockheed Martin's new streamlined procedures, VIF operations began about nine days before workers rolled the vehicle out to the pad. The ATLAS V was ready to go when it departed the VIF, and the vehicle could be launched a mere 12 hours later. Since most of the ATLAS V Aerospace Ground Equipment (AGE) resided in the VIF and the ATLAS V Spaceflight Operations Center, very little ground equipment was exposed to blast damage in the event of a launch mishap or accident on the pad. The new design promoted safety, efficiency and flexibility.

210. The four-story-tall ASOC was located four miles from the launch pad. It replaced 13 old ATLAS facilities, and it gathered customer support, vehicle checkout, and launch control operations under one roof. The ASOC included a two-story amphitheater, a Mission Operations Center, a two-story launch control center, and various management/engineering support rooms. With 30,000 square feet of floor space, the ASOC could process up to six ATLAS Vs at a time. Since most checkout operations would be completed at Lockheed Martin's Denver plant before the ATLAS V was shipped, checkout at the ASOC might require as little as one day of work for each vehicle.

211. 45th Space Wing History, 1 Jan –31 Dec 98, Vol I, p 50; 45th Space Wing History, 1 Jan – 31 Dec 2000, Vol I, p 23; 45th Space Wing History, 1 Jan – 31 Dec 01, Vol I, p 28; "Lockmart Bets Launch Future on Atlas V," *Aviation Week & Space Technology,* 10 Dec 01, pp 64, 65; E-Mail, Suzanne Schulman, 45 CES/CEL to 45SWHO, "Input for Cape History Books," 11 Jan 2000; Discussion, M. Cleary, 45 SW/HO, with Ms. Suzanne Schulman, 45 CES/CEL, 8 Feb 2000.

212. "Lockmart Bets Launch Future on Atlas V," *Aviation Week & Space Technology*, 10 Dec 01, pp 64, 65; "Lockheed Martin Ships Atlas 5 Segment to NASA," *Space News*, 14 May 01, p 14; "Atlas 5 arrives for first flight," *Florida Today*, 7 Jun 01, p 1B; "Atlas V booster arrives," *45th Space Wing Missileer*, 8 Jun 01, p 9; "Atlas V Stacked," *Aviation Week & Space Technology*, 29 Oct 01, p 27; Discussion, M. Cleary, 45 SW/HO with Mr. Ken Warren, 45 SW/PA, 19 Feb 02; Summary, Lockheed Martin Space Systems, "Atlas V/EELV Chronology," undated.

213. The ATLAS V's Aerojet SRMs were about 61 inches in diameter and 67 feet long. Each one provided about 306,000 pounds of thrust. All SRMs ignited at lift-off, and they burned for about 95 seconds. Beginning around 106 seconds into the flight, the first two SRMs was jettisoned, followed by the remaining SRM two-seconds later. Depending on the configuration of the ATLAS V booster and number of SRMs, the vehicle was able to lift payloads ranging from 10,900 pounds to 19,180 pounds to Geosynchronous Transfer Orbit.

214. 45 SW History, 1 Jan -31 Dec 06, Vol I, p 64.

215. *Ibid.*

216. 45 SW History, 1 Jan -31 Dec 06, Vol I, pp 64, 65.

217. 45 SW History, 1 Jan -31 Dec 06, Vol I, p 65.

218. 45 SW History, 1 Jan – 31 Dec 02, Vol I, pp 95, 96; Summary, International Launch Services, "HOT BIRD 6 Mission Overview," Aug 02; New Release, International Launch Services, "Inaugural Atlas V Scores Success for ILS, Lockheed Martin," 4 Sep 02.

219. Astrotech began processing payloads for the Cape in 1985, but the HOT BIRD 6 was the first payload to be processed in the Lee facility. Astrotech named the facility after the late Captain Chet Lee (former president of SpaceHab and chairman of Astrotech's board of directors). The 50,000-square-foot facility was 124 feet tall, and it was designed specifically to process spacecraft and integrate payload fairings associated with ATLAS V and DELTA IV space missions.

220. 45 SW History, 1 Jan – 31 Dec 02, Vol I, p 96; Summary, Lockheed Martin Space Systems, "Atlas V/EELV Chronology," undated; "New EELV Launch Ops Spur Astrotech Growth," *Aviation Week & Space Technology*, 12 Aug 02.

221. 45 SW History, 1 Jan – 31 Dec 02, Vol I, p 97; "Atlas 5 ready for first launch today," *Florida Today*, 21 Aug 02, p 1A; Fact Sheet, Lockheed Martin, "Atlas V," 4 Sep 02; "Lockheed prepares for next mission," *Florida Today*, 23 Aug 02, p 3B; "Atlas V Soars, Market Slumps," *Aviation Week & Space Technology*, 26 Aug 02, p 22; "Atlas V Flies, But Will It Sell?" *Aviation Week & Space Technology*, 16 Sep 02, p 22; "Wing helps launch new space era," *45th Space Wing Missileer*, 16 Aug 02, p 1; "Atlas V rumbles to success," *45th Space Wing Missileer*, 23 Aug 02, p 1.

222. 45 SW History, 1 Jan – 31 Dec 03, Vol I, p 86; New Release, International Launch Services, "ILS Launches Hellas-Sat on Atlas V," 16 Sep 03.

223. The first instrumentation problem involved an Advanced Transportable Optical Tracking System (ATOTS) at one of the range's camera sites. The ATOTS was Partially Mission Capable (PMC) due to intermittent "No-Go" alarms at L minus 329 minutes. Engineers isolated the problem to a failed time code generator. They adjusted the time code generator, and the problem cleared. Consequently, officials declared the ATOTS Fully Mission Capable before the launch. The other problem involved a radar. Officials declared the radar Non-Mission Capable (NMC) at L minus 216 minutes "due to no video being received in the [Range Operations Control Center]." The radar was repaired, and it supported the rest of the countdown "nominally."

224. 45 SW History, 1 Jan – 31 Dec 03, Vol I, pp 86, 88; News Release, International Launch Services, "Successful Launch of the Atlas V for Hellas-Sat," 13 May 03.

225. ILS was formed in 1995 as a joint venture by Lockheed Martin and its two Russian partners — RSC Energia and Khrunichev State Research and Production Space Center. Headquartered in McLean, Virginia, ILS marketed and managed missions involving the Lockheed-Martin ATLAS and Russian-built PROTON and ANGARA space launch vehicles.

226. The 500 series vehicle chosen for the mission was designated '521' to indicate it would carry a '5' meter fairing, '2' Aerojet solid rocket motors, and '1' RL10A-4-2 engine in its CENTAUR upper stage.

227. 45 SW History, 1 Jan – 31 Dec 03, Vol I, p 88; News Release, International Launch Services, "ILS Launches Rainbow 1 Satellite," 16 Sep 03.

228. 45 SW History, 1 Jan – 31 Dec 03, Vol I, p 89; News Release, International Launch Services, "ILS Launches Rainbow 1 Satellite," 16 Sep 03.

229. 45 SW History, 1 Jan – 31 Dec 04, Vol I, p 84.

230. *Ibid.*

231. 45 SW History, 1 Jan – 31 Dec 04, Vol I, p 85.

232. *Ibid.*

233. The AV-004 mission was actually the fifth ATLAS V mission launched from Cape Canaveral. AV-005 (the fourth ATLAS V mission) had been launched from the Cape on 17 December 2004.

234. According to Inmarsat's plans for the near future, the first INMARSAT-4 would be complemented with another INMARSAT-4 (F2) to provide services to mobile users in South America, most of North America, in the Atlantic, and (part of) the Pacific. It would be launched aboard a Sea Launch (company) rocket. A third INMARSAT-4 (F3) was built as a backup for the two-satellite constellation. It might also serve as an additional operational satellite over the Pacific Ocean.

235. 45 SW History, 1 Jan – 31 Dec 05, Vol I, p 76.

236. *Ibid.*

237. 45 SW History, 1 Jan – 31 Dec 05, Vol I, p 77

238. *Ibid.*

239. 45 SW History, 1 Jan – 31 Dec 06, Vol I, p 69.

240. *Ibid.*

241. 45 SW History, 1 Jan – 31 Dec 06, Vol I, pp 69, 70.

242. 45 SW History, 1 Jan – 31 Dec 06, Vol I, p 70.

243. 45 SW History, 1 Jan -31 Dec 05, Vol I, p 80.

244. *Ibid.*

245. 45 SW History, 1 Jan -31 Dec 05, Vol I, pp 80, 81.

246. 45 SW History, 1 Jan -31 Dec 05, Vol I, p 81

247. 45 SW History, 1 Jan -31 Dec 06, Vol I, p 66.

248. *Ibid.*

249. 45 SW History, 1 Jan -31 Dec 06, Vol I, pp 66, 67.

250. 45 SW History, 1 Jan -31 Dec 06, Vol I, p 67.

251. "A Historical Look at United States Launch Vehicles, 1967 - Present," ANSER's International Aerospace Division (1994), pp II.C-12, II.C-18, II.C-19; Excerpt, ANSER, "1996 Space Launch Activities," Jan 97, p III-26; 45 SW History, 1 Jan – 31 Dec 98, Vol I, p 70; Fact Sheet, Headquarters Air Force Space Command, "Titan IVB," Apr 05.

252. "A Historical Look at United States Launch Vehicles, 1967 - Present," ANSER's International Aerospace Division (1994), p II.C-18; Excerpt, ANSER, "1996 Space Launch Activities," Jan 97, p III-26; Fact Sheet, Lockheed Martin, "TITAN, America's Silent Hero," undated; 45 SW History, 1 Jan – 31 Dec 97, Vol I, pp 71, 72; 45 SW History, 1 Jan – 31 Dec 98, Vol I, pp 70, 71.

253. "A Historical Look at United States Launch Vehicles, 1967 - Present," ANSER's International Aerospace Division (1994), pp II.C-17, II.C-19; Fact Sheet, Lockheed Martin, "TITAN, America's Silent Hero," undated; 45 SW History, 1 Jan – 31 Dec 97, Vol I, pp 71, 72; 45 SW History, 1 Jan – 31 Dec 98, Vol I, pp 70, 71.

254. Excerpt, ANSER, "1996 Space Launch Activities," Jan 97, p III-26; Fact Sheet, Lockheed Martin, "TITAN, America's Silent Hero," undated; 45 SW History, 1 Jan – 31 Dec 97, Vol I, pp 71.

255. Fact Sheet, Lockheed Martin, "TITAN, America's Silent Hero," undated; 45 SW History, 1 Jan – 31 Dec 98, Vol I, pp 71, 72; 45 SW History, 1 Jan -31 Dec 05, Vol I, p 50.

256. 45 SW History, 1 Jan – 31 Dec 98, Vol I, pp 71, 72; 45 SW History, 1 Jan -31 Dec 05, Vol I, p 50.

257. 45 SW History, 1 Jan – 31 Dec 98, Vol I, p 72.

258. 45 SW History, 1 Jan – 31 Dec 98, Vol I, pp 72, 73.

259. The MILSTAR was a communications satellite designed to provide secure, jam-resistant communications between command authorities and ground forces, ships, submarines, and aircraft worldwide. The spacecraft weighed approximately 10,000 pounds, and it measured 51 feet long and 116 feet wide with its solar panels extended. (SOURCE: *Air Force Magazine,* "2001 USAF Almanac," May 01, p 157.)

260. The launch failure at Vandenberg occurred approximately 101 seconds after lift-off. Speculating that the mission may have cost as much as $1 billion, the media characterized the mishap as "the second most expensive space disaster since the Challenger tragedy." A subsequent investigation of the mishap revealed a flawed SRM segment caused the flight failure. The TITAN IV fleet was grounded pending thorough examination of all SRMs, and workers accomplished TITAN IV inspections and stacking operations under the closest scrutiny in the months that followed.

261. Summary, CSR, "Range Operations Summary, TITAN IV/MILSTAR Launch, TC-12," Jan 94, p 6; 45 SW History, 1 Jan – 31 Dec 94, Vol I, p 83.

262. 45 SW History, 1 Jan – 31 Dec 94, Vol I, p 84.

263. Summary, SMC/MEFC, "K-7 Top Issues/Major Events," 3 May 94; Excerpt, Martin Marietta Space Launch Systems, "TITAN IV-7 Post-Flight Report," 17 Jun 94, p 5-6; "New delay plagues Titan 4 program," *Florida Today,* 17 Feb 93.

264. Summary, SMC/MEFC, "K-7 Top Issues/Major Events," 3 May 93; Excerpt. Martin Marietta Space Launch Systems, "TITAN IV-9 Post-Flight Report," 10 Oct 94, p 5-6.

265. "New delay plagues Titan 4 program," *Florida Today,* 17 Feb 93; Summary, SMC/MEFC, "K-7 Top Issues/Major Events," 3 May 93; Excerpt, Martin Marietta Space Launch Systems, "TITAN IV-7 Post Flight Report," 17 Jun 94, p 5-6; Input, SMC/MEFC, "CY 1994 History Coordination," ca. 16 Jun 95.

266. Discussion, M. Cleary, 45 SW/HO with Mr. Dan Berlinrut, 45 SW/SESL, 2 Feb 95.

267. 45 SW History, 1 Jan – 31 Dec 94, Vol I, p 87; Excerpt, Martin Marietta Space Launch Systems, "TITAN IV-7 Post-Flight Report," 17 Jun 94, pp 5-1, 5-2, 3-1, 3-2, 3-4 and 3-5; "Battery problem forces Titan IV scrub," *45th Space Wing Missileer,* 29 Apr 94; "Titan roars into space," *45th Space Wing Missileer,* 6 May 94.

268. Summary, Captain Jacobs, 5 SLS "TIV-9 Processing History Summary," 21 Sep 94; Excerpt, Martin Marietta Space Launch Systems, "TITAN IV-9 Post-Flight Report," 10 Oct 94, p 5-6; Summary, Captain Eric Spittle, SMC/MEFC, "K-9 Top Issues/Major Events," 27 Aug 94.

269. *Ibid.*

270. See note 268.

271. 45 SW History, 1 Jan – 31 Dec 94, Vol I, p 89.

272. 45 SW History, 1 Jan – 31 Dec 94, Vol I, p 90.

273. 45 SW History, 1 Jan – 31 Dec 94, Vol I, p 90; CSR, "Telemetry Performance Evaluation Report, Operation D9646 - TITAN IV/CENTAUR," 7 Oct 94; Summary, 45 RANS/DS, "Major Operations FY 94," 25 Oct 94, p 3.

274. The Defense Support Program spacecraft was an early warning satellite designed to alert officials of a possible ballistic missile attack on the United States or its military forces. The DSP spacecraft weighed about 5,000 pounds, and it measured 22 feet in diameter and 32 feet 9 inches wide with its solar panels extended. (SOURCE: *Air Force Magazine*, "2001 USAF Almanac," May 01, p 157.)

275. 45 SW History, 1 Jan - 31 Dec 94, Vol I, p 91.

276. 45 SW History, 1 Jan - 31 Dec 94, Vol I, p 92.

277. "Air Force Launches TITAN IV," *45th Space Wing Missileer,* 13 Jan 95, p 1; CSR, "Telemetry Performance Evaluation Report, Operation E7131 - TITAN IV/IUS," 25 Jan 95, p 3; 45 SW History, 1 Jan - 31 Dec 94, Vol I, p 93.

278. Excerpt, Martin Marietta Technologies, Inc., "TITAN IV-19 Post Flight Report," 24 Aug 95, p 5-7.

279. Summary, M. Cleary, 45 SW/HO, "Interviews of 4 Jan 96," 4 Jan 96; "Air Force launches TITAN IV/CENTAUR," 19 May 95, pp 1, 3; History Coordination, 5 SLS/DOU, 15 May 96.

280. Excerpt, Martin Marietta Rapid Post-Flight Team, "TITAN IV-23, Launch/Flight Directive Post-Flight Mission Review Report," 26 May 95; 45 SW History, 1 Jan – 31 Dec 95, Vol I, pp 85, 86.

281. Excerpt, Martin Marietta Technologies, Inc., "TITAN IV-19 Post Flight Report," 24 Aug 95, pp 5-1, 5-7.

282. 45 SW History, 1 Jan – 31 Dec 95, Vol I, pp 86, 88.

283. Excerpt, Martin Marietta Technologies, Inc., "TITAN IV-21 Post Flight Report," 19 Dec 95, pp 5-1, 5-8; 45 SW History, 1 Jan- 31 Dec 95, pp 88, 89.

284. 45 SW History, 1 Jan- 31 Dec 95, p 90.

285. "Titan IV carries Milstar into clear skies," *45th Space Wing Missileer*, 10 Nov 95, p 1; Summary, 45 RANS/ DOUS, "Major Operations FY-96," 23 Feb 96; Excerpt, Martin Marietta Technologies, Inc., "TITAN IV-21 Post Flight Report," 19 Dec 95, pp 3-1, 3-2; E-Mail, AFNS, "Crosslink marks space milestone," 24 Jan 96.

286. Excerpt, Lockheed Martin Corporation, "TITAN IV-2 Post-Flight Report," 15 Aug 96, p 5-13.

287. Excerpt, Lockheed Martin Corporation, "TITAN IV-2 Post-Flight Report," 15 Aug 96, pp 8-3, 8-7, 8-9, 8-11; 45 SW History, 1 Jan – 31 Dec 96, Vol I, p 76; Summary, 45 RANS/DOUS, "Major Operations FY-96," 9 Jul 96, p 4.

288. Excerpt, Lockheed Martin Corporation, "TITAN IV-2 Post Flight Report," 15 Aug 96, p 5-10.

289. 45 SW History, 1 Jan- 31 Dec 96, Vol I, p 77; "TITAN goes up in blaze of glory," *45th Space Wing Missileer*, 12 Jul 96, p 1.

290. Among the dead were five members of the 71st Rescue Squadron, assigned to Patrick AFB. They were Captain Christopher J. Adams, Captain Leland T. Haun, MSgt Michael G. Heiser, SSgt Kevin Jerome Johnson, and A1C Justin R. Wood. They had been deployed to King Abdul Aziz Air Base near Dhahran as HC-130P aircraft crewmembers. Fifty-one people from Patrick AFB had been deployed to Dhahran on various support missions at the time of the bombing. In addition to the dead, 14 of Patrick's people were among 547 injured in the blast.

291. 45 SW History, 1 Jan- 31 Dec 96, Vol I, p 77; "TITAN goes up in blaze of glory," *45th Space Wing Missileer*, 12 Jul 96, p 1; "Warnings plentiful in Saudi bombing," *USA Today*, 26 Aug 96, p 16A; "For those who died defending freedom," *45th Space Wing Missileer* (Memorial Issue), 5 Jul 96, pp 7-9; "Five Patrick airmen killed, more wounded in explosion," *45th Space*

Wing Missileer, 28 Jun 96; "Relatives await return of the wounded," 28 Jun 96, p 4A; "Patrick dealt another blow," *Florida Today*, 28 Jun 96, pp 1, 5A; "Local services will remember the slain," *Florida Today*, 28 Jun 96, p 4A.

292. The TITAN IV "A" model was just as tall as the new "B" model, but each of the SRMUs on the TITAN IVB were six inches wider than the TITAN IVA's solid rocket motors (SRMs). Due to the increased width of the vehicle, all 20 levels of the Mobile Service Tower received narrower platforms. Unlike the basic TITAN IVA vehicle, the TITAN IVB came to the launch pad with its SRMUs fully stacked. In addition to those hardware and procedural changes, TITAN IV launch operations software was upgraded to Programmable Aerospace Ground Equipment (PAGE) for the launch pads, operations center, and ancillary control centers.

293. Launch Book, CSR, "Launch Book, TITAN IVB, Mission B-24/IUS-4," Feb 97, p A-1; Excerpt, Lockheed Martin Corporation, "TITAN IVB-24 Post Flight Report," 8 Apr 97, pp 4-1, 4-11, 5-4; "A New Year & A New Rocket," *45th Space Wing Missileer*, undated; "Titan IVB team ready for launch," *45th Space Wing Missileer*, 14 Feb 97, pp 1, 4; "First-ever Titan IVB launch a success," *45th Space Wing Missileer*, 28 Feb 97, p 11.

294. Excerpt, Lockheed Martin Corporation, "TITAN IVB-24 Post Flight Report," 8 Apr 97, pp 3-1, 3-2, 3-3, 3-9; "First-ever Titan IVB launch a success," *45th Space Wing Missileer*, 28 Feb 97, p 11; 45 SW History, 1 Jan –31 Dec 97, Vol I, pp 75, 76.

295. 45 SW History, 1 Jan – 31 Dec 97, Vol I, pp 79, 81.

296. Excerpt, Lockheed Martin Corporation, "TITAN IVA-17 Post Flight Report," 19 Dec 97, pp 3-1, 3-2, 3-3; "Air Force sets Titan launch record." *45th Space Wing Missileer*, 14 Nov 97, pp 1, 5; 45 SW History, 1 Jan – 31 Dec 97, Vol I, p 81.

297. 45 SW History, 1 Jan-31 Dec 98, Vol I, p 73.

298. Excerpt, Lockheed Martin, "TITAN IVB-25 Post Flight Report," 19 Jun 98, pp 3-1, 3-2, 3-3, 3-11; 45 SW History, 1 Jan-31 Dec 98, Vol I, pp 73, 74.

299. 45 SW History, 1 Jan - 31 Dec 98, Vol I, pp 74, 76.

300. 45 SW History, 1 Jan - 31 Dec 98, Vol I, p 76.

301. 45 SW History, 1 Jan - 31 Dec 98, Vol I, pp 76, 77; News Release, 45 SW/PA, "Titan Explodes in Flight," 12 Aug 98; News Release, 45 SW/PA, "Titan Explosion Update," 12 Aug

98; "Training pays off during Titan mishap," *45th Space Wing Missileer*, 14 Aug 98, pp 1, 4; "A major malfunction - Explosion turns spy satellite, rocket into billion-dollar fireworks show," *Florida Today,* 13 Aug 98, p 1; "Major malfunction destroys rocket, satellite," *Florida Today,* 13 Aug 98, p 12A.

302. 45 SW History, 1 Jan - 31 Dec 98, Vol I, p 77; "Beachgoers, boaters warned to stay clear of debris," *Florida Today,* 13 Aug 98; "New emergency plan pays off for Brevard," *Florida Today,* 13 Aug 98, p 7A.

303. Excerpt, HQ AFSPC, "Investigation of USAF Launch Vehicle Accident, Titan IV A-20, 12 August 1998," undated, pp 5, 13, 15, 16, 17; "Titan mishap investigation findings released," *45th Space Wing Missileer,* 22 Jan 99, p 7.

304. 45 SW History, 1 Jan – 31 Dec 99. p 60; *Air Force Magazine*, "2001 USAF Almanac," May 01, p 157.

305. 45 SW History, 1 Jan – 31 Dec 99. p 60.

306. 45 SW History, 1 Jan – 31 Dec 99. pp 60, 61.

307. 45 SW History, 1 Jan – 31 Dec 99. p 61; "Titan launches busy month in April," *45th Space Wing Missileer,* 16 Apr 99, p 1; "Separation system flaw blamed for launch failure," *Florida Today* Space Online, 18 Aug 99.

308. 45 SW History, 1 Jan – 31 Dec 99, Vol I, p 63.

309. 45 SW History, 1 Jan – 31 Dec 99, Vol I, p 63; "Titan, Delta Failures Force Sweeping Reviews," *Aviation Week & Space Technology,* 10 May 99; "Air Force says Milstar satellite dead," *Florida Today* Space Online, 13 May 99.

310. "Titan, Delta Failures Force Sweeping Reviews," *Aviation Week & Space Technology,* 10 May 99; "Panel Links Launch Failures to Systemic Ills," *Aviation Week & Space Technology,* 13 Sep 99, pp 24, 25; "Lockheed Martin Implements space panel recommendations," *Florida Today* Space Online, 9 Sep 99.

311. 45 SW History, 1 Jan – 31 Dec 2000, pp 84, 85.

312. 45 SW History, 1 Jan – 31 Dec 2000, p 85; "Titan returns to space in grand style," *45th Space Wing Missileer,* 12 May 00, p 1.

313. 45 SW History, 1 Jan – 31 Dec 01, Vol I, pp 78, 79, 80; "Milstar II To Propel New Defense Info Infrastructure," *Aviation Week & Space Technology*, 5 Mar 01, pp 42, 43.

314. The MST could not be moved around the vehicle until the drive system was repaired. Troubleshooting was completed on 17 June, and the vehicle was moved to the pad on 19 June 2000. The CENTAUR upper stage was mated to the vehicle on 23 June 2000.

315. 45 SW History, 1 Jan – 31 Dec 01, Vol I, pp 78, 78, 80.

316. A water leak drained the pump station's reservoir, and several pumps were damaged when the emergency shut-off failed. The incident delayed propellant loading operations by one day.

317. 45 SW History, 1 Jan – 31 Dec 01, Vol I, p 80.

318. 45 SW History, 1 Jan – 31 Dec 01, Vol I, p 81; "MILSTAR soars on Titan," *45th Space Wing Missileer*, 2 Mar 01, p 1.

319. "Titan IVB, Delta Missions Near $1 Billion in Operations," *Aviation Week & Space Technology*, 30 Jul 01, p 31; 45 SW History, 1 Jan – 31 Dec 01, Vol I, p 83.

320. 45 SW History, 1 Jan – 31 Dec 01, Vol I, pp 83, 84; "Hardware Problems Delay Genesis, DSP," *Aviation Week & Space Technology*, 6 Aug 01.

321. 45 SW History, 1 Jan – 31 Dec 01, Vol I, p 85.

322. 45 SW History, 1 Jan –31 Dec 02, Vol I, pp 76, 77; Milstar Ring To Speed Data Toward Combat Zones," *Aviation Week & Space Technology*, 21 Jan 02, p 28; "U.S. Military Wants Sweeping Satcom Changes," *Aviation Week & Space Technology*, 21 Jan 02, pp 27, 28.

323. 45 SW History, 1 Jan –31 Dec 02, Vol I, pp 77, 78.

324. 45 SW History, 1 Jan –31 Dec 02, Vol I, p 78.

325. 45 SW History, 1 Jan –31 Dec 02, Vol I, pp 79, 80.

326. 45 SW History, 1 Jan –31 Dec 02, Vol I, p 80; "TITAN IV launch success," *45th Space Wing Missileer*, 18 Jan 02, p 1; "Military satellite in orbit, functional, Air Force says," *Florida Today*, 17 Jan 02, p 1B.

327. 45 SW History, 1 Jan – 31 Dec 03, Vol I, p 63; *Air Force Magazine*, "2001 USAF Almanac," May 01, p 157; "Milstar to Speed Communications With Afghanistan," *Space News*, 21 Jan 02, p 10; "Milstar Ring To Speed Data Toward Combat Zones," *Aviation Week & Space Technology*, 21 Jan 02, p 28; "U.S. Military Wants Sweeping Satcom Changes," *Aviation Week & Space Technology*, 21 Jan 02, pp 27, 28; "Wing launches vital to Operation Iraqi Freedom," *45th Space Wing Missileer*, 4 Apr 03, p 1; "Launch provides warfighter support," *45th Space Wing Missileer*, 11 Apr 03, p 1.

328. A new consignment of FTS antennas arrived at the Cape around 26 November 2002. They were installed by 9 December, and RF Link analysis was accomplished with good results. At Range Safety's insistence, Lockheed Martin continued to test the FTS antennas each week to ensure they would be ready to support the launch. The issue was finally closed when Range Safety approved the FTS antennas for flight in early April 2003.

329. Lockheed Martin completed battery testing on 17 December 2002. Following a data review on 19 December 2002, Lockheed Martin and Range Safety gave their final approval to use the five amp-hour batteries on the MILSTAR (Flight 6) mission.

330. A shipment of six new 250 amp-hour CENTAUR batteries arrived at the Cape on 22 November 2002. The Mission Risk Decision Board decided on a battery configuration that offered the "greatest performance margin" on 5 December 2002. Officials closed out the matter on 6 December 2002.

331. 45 SW History, 1 Jan – 31 Dec 03, pp 64, 65.

332. 45 SW History, 1 Jan – 31 Dec 03, pp 65, 66.

333. 45 SW History, 1 Jan – 31 Dec 03, pp 66, 67.

334. 45 SW History, 1 Jan – 31 Dec 03, pp 67, 68; "Launch provides warfighter support," *45th Space Wing Missileer*, 11 Apr 03, p 1.

335. 45 SW History, 1 Jan – 31 Dec 03, Vol I, pp 68, 70.

336. 45 SW History, 1 Jan – 31 Dec 03, Vol I, p 70.

337. 45 SW History, 1 Jan – 31 Dec 03, Vol I, pp 71, 72.

338. 45 SW History, 1 Jan – 31 Dec 03, Vol I, pp 72.

339. 45 SW History, 1 Jan – 31 Dec 03, Vol I, p 73.

340. Early analysis of the failure indicated that a cooling line malfunction probably caused "bearing grinding" inside the pump's motor, and this condition degenerated to the point that oxidizer seeped into the motor windings with explosive consequences for the pump. Officials closed out their investigation on 18 August 2003.

341. 45 SW History, 1 Jan – 31 Dec 03, Vol I, p 74; "Straight from the Commander's Desk," *45th Space Wing Missileer*, 22 Aug 03, p 2.

342. 45 SW History, 1 Jan – 31 Dec 03, Vol I, p 75.

343. Major Homer was the first officer aboard United Airlines Flight 93, which crashed near Shanksville, Pennsylvania on 11 September 2001. Major Homer served in Desert Shield and Desert Storm, and he was considered "one of the heroes of 9/11."

344. 45 SW History, 1 Jan – 31 Dec 03, Vol I, pp 75, 76; "Titan mission dedicated to AF major lost on 9/11," *45th Space Wing Missileer*, 12 Sep 03, p 3; "A Titan IVB/Centaur launch vehicle lifts off from Complex 40…," *Aviation Week & Space Technology*, 15 Sep 03, p 21.

345. 45 SW History, 1 Jan -31 Dec 04, Vol I, pp 63, 64.

346. 45 SW History, 1 Jan -31 Dec 04, Vol I, pp 64, 65.

347. 45 SW History, 1 Jan -31 Dec 04, Vol I, p 65.

348. 45 SW History, 1 Jan -31 Dec 05, Vol I, p 63.

349. *Ibid.*

350. 45 SW History, 1 Jan -31 Dec 05, Vol I, p 64.

351. *Ibid.*

352. See Note 350.

353. 45 SW History, 1 Jan -31 Dec 05, Vol I, p 65.

354. 45 SW History, 1 Jan -31 Dec 05, Vol I, pp 65, 66.

355. 45 SW History, 1 Jan -31 Dec 05, Vol I, p 66.

356. The CASSINI spacecraft was named in honor of 17th Century Italian astronomer Gian Domenico Cassini, who discovered the largest gap in Saturn's ring system. CASSINI was a joint U.S.-European mission to Saturn and its moon Titan. After its launch in October 1997, the 12,670-pound spacecraft employed a Venus-Venus-Earth-Jupiter gravity assist flight trajectory to arrive at a Saturn orbit on 1 July 2004. NASA expected the spacecraft to spend four years surveying the planet, its ring system, and moons. CASSINI released the Huygens Probe for its two and one-half hour descent through Titan's atmosphere on 14 January 2005. NASA's Jet Propulsion Laboratory (JPL) in Pasadena, California, managed the CASSINI project.

357. Launch Book, CSR, "Launch Book, TITAN IVB, B-33/K-33/TC-21 Mission," Oct 97, pp A-1, A-4; Summary, NASA, "CASSINI Fact Sheet," undated, p 1; "CASSINI Propulsion System To Be Delivered in August," *Aviation Week & Space Technology*, 15 Jul 96, pp 55; "CASSINI gliding smoothly on Saturn voyage," *Florida Today,* 17 Oct 97, p 2; Excerpt, Lockheed Martin Corporation, "Titan IVB-33 Post-Flight Report," 26 Nov 97, pp 1-1, 2-1, 3-1, 4-11, 5-1; "Cassini successfully arrives at Saturn," William Harwood, CBS News, 1 Jul 04; Summary, European Space Agency (ESA), "Cassini/Huygens Mission Facts," 26 Apr 05.

358. Fax, Jeff Dobbins, Department of Justice, to Sara Najjar-Wilson, "Order Denying Plaintiffs' Motion for Temporary Restraining Order and Plaintiffs' Motion for Preliminary Injunction," 11 Oct 97; "Anti-CASSINI group plans two more rallies before liftoff," *Florida Today,* 10 Oct 97, p 5A.

359. 45 SW History, 1 Jan – 31 Dec 97, Vol I, p 77.

360. 45 SW History, 1 Jan – 31 Dec 97, Vol I, p 79; "CASSINI gliding smoothly on Saturn voyage," *Florida Today,* 17 Oct 97, p 2; Excerpt, Lockheed Martin Corporation, "Titan IVB-33 Post-Flight Report," 26 Nov 97, pp 3-1, 3-2, 3-3, 3-12, 3-13.

ATLAS AND TITAN LAUNCH SYNOPSIS

1 January 1993 – 31 December 2006

25 March 1993	ATLAS I/CENTAUR	Pad 36B	UHF F/O #1
19 July 1993	ATLAS II/CENTAUR	Pad 36A	DSCS III
3 September 1993	ATLAS I/CENTAUR	Pad 36B	UHF F/O #2
28 November 1993	ATLAS II/CENTAUR	Pad 36A	DSCS III
16 December 1993	ATLAS IIAS/CENTAUR (1st ATLAS IIAS Flt)	Pad 36B	TELSTAR 4
7 February 1994	TITAN IVA/CENTAUR	Pad 40	MILSTAR (Flt 1)
13 April 1994	ATLAS I/CENTAUR	Pad 36B	GOES-I
3 May 1994	TITAN IVA/CENTAUR	Pad 41	Classified Payload
24 June 1994	ATLAS I/CENTAUR	Pad 36B	UHF F/O #3
3 August 1994	ATLAS IIA/CENTAUR	Pad 36A	DIRECTV
27 August 1994	TITAN IVA/CENTAUR	Pad 41	Classified Payload
6 October 1994	ATLAS IIAS/CENTAUR	Pad 36B	INTELSAT VII
29 November 1994	ATLAS IIA/CENTAUR	Pad 36A	ORION
22 December 1994	TITAN IVA/IUS	Pad 40	DSP
10 January 1995	ATLAS IIAS/CENTAUR	Pad 36B	INTELSAT VII
29 January 1995	ATLAS II/CENTAUR	Pad 36A	EHF F/O#4
22 March 1995	ATLAS IIAS/CENTAUR	Pad 36B	INTELSAT VII
7 April 1995	ATLAS IIA/CENTAUR	Pad 36A	MSAT
14 May 1995	TITAN IVA/CENTAUR	Pad 40	Classified Payload
23 May 1995	ATLAS I/CENTAUR	Pad 36B	GOES-J
31 May 1995	ATLAS II/CENTAUR	Pad 36A	EHF F/O#5
10 July 1995	TITAN IVA/CENTAUR	Pad 41	Classified Payload
31 July 1995	ATLAS IIA/CENTAUR	Pad 36A	DSCS III
29 August 1995	ATLAS IIAS/CENTAUR	Pad 36B	JCSAT
22 October 1995	ATLAS II/CENTAUR	Pad 36A	EHF F/O#6
6 November 1995	TITAN IVA/CENTAUR	Pad 40	MILSTAR (Flt 2)
2 December 1995	ATLAS IIAS/CENTAUR	Pad 36B	SOHO
15 December 1995	ATLAS IIA/CENTAUR	Pad 36A	GALAXY IIIR
1 February 1996	ATLAS IIAS/CENTAUR	Pad 36B	PALAPA-C1
3 April 1996	ATLAS IIA/CENTAUR	Pad 36A	INMARSAT-3
24 April 1996	TITAN IVA/CENTAUR	Pad 41	Classified Payload
30 April 1996	ATLAS I/CENTAUR	Pad 36B	SAX

3 July 1996	TITAN IVA/NUS	Pad 40	Classified Payload
25 July 1996	ATLAS II/CENTAUR	Pad 36A	UHF F/O#7
8 September 1996	ATLAS IIA/CENTAUR	Pad 36B	GE-1
21 November 1996	ATLAS IIA/CENTAUR	Pad 36B	HOT BIRD 2
18 December 1996	ATLAS IIA/CENTAUR	Pad 36A	INMARSAT-3 F3
17 February 1997	ATLAS IIAS/CENTAUR	Pad 36B	JCSAT-4
23 February 1997	TITAN IVB/IUS (1st TITAN IVB Launch)	Pad 40	DSP
8 March 1997	ATLAS IIA/CENTAUR	Pad 36A	TEMPO II
25 April 1997	ATLAS I/CENTAUR	Pad 36B	GOES-K
28 July 1997	ATLAS IIAS/CENTAUR	Pad 36B	SUPERBIRD-C
4 September 1997	ATLAS IIAS/CENTAUR	Pad 36A	GE-3
5 October 1997	ATLAS IIAS/CENTAUR	Pad 36B	ECHOSTAR III
15 October 1997	TITAN IVB/CENTAUR	Pad 40	CASSINI
25 October 1997	ATLAS IIA/CENTAUR	Pad 36A	DSCS III
8 November 1997	TITAN IVA/CENTAUR	Pad 41	Classified Payload
8 December 1997	ATLAS IIAS/CENTAUR	Pad 36B	GALAXY VIII-i
29 January 1998	ATLAS IIA/CENTAUR	Pad 36A	NRO Payload
28 February 1998	ATLAS IIAS/CENTAUR	Pad 36B	INTELSAT 806
16 March 1998	ATLAS II/CENTAUR	Pad 36A	UHF F-8
9 May 1998	TITAN IVB/CENTAUR	Pad 40	NRO Payload
18 June 1998	ATLAS IIAS/CENTAUR	Pad 36A	INTELSAT 805
12 August 1998	TITAN IVA/CENTAUR	Pad 41	NRO Payload
9 October 1998	ATLAS IIA/CENTAUR	Pad 36B	HOTBIRD-5
20 October 1998	ATLAS IIA/CENTAUR	Pad 36A	UHF F-9
16 February 1999	ATLAS IIAS/CENTAUR	Pad 36A	JCSAT-6
9 April 1999	TITAN IVB/IUS	Pad 41	DSP-19
12 April 1999	ATLAS IIAS/CENTAUR	Pad 36A	EUTELSAT W3
30 April 1999	TITAN IVB/CENTAUR	Pad 40	MILSTAR (Flt 3)
23 September 99	ATLAS IIAS/CENTAUR	Pad 36A	ECHOSTAR V
23 November 1999	ATLAS IIA/CENTAUR	Pad 36B	UHF F-10
21 January 2000	ATLAS IIA/CENTAUR	Pad 36A	DSCS III B8
3 February 2000	ATLAS IIAS/CENTAUR	Pad 36B	HISPASAT 1-C
3 May 2000	ATLAS IIA/CENTAUR	Pad 36A	GOES-L
8 May 2000	TITAN IVB/IUS	Pad 40	DSP-20
24 May 2000	ATLAS IIIA/CENTAUR (1st ATLAS IIIA Launch)	Pad 36B	EUTELSAT W4

30 June 2000	ATLAS IIA/CENTAUR	Pad 36A	TDRS-H
14 July 2000	ATLAS IIAS/CENTAUR	Pad 36B	ECHOSTAR VI
20 October 2000	ATLAS IIA/CENTAUR	Pad 36A	DSCS III B11
6 December 2000	ATLAS IIAS/CENTAUR	Pad 36A	NRO MLV-11
27 February 2001	TITAN IVB/CENTAUR	Pad 40	MILSTAR (Flt 4)
19 June 2001	ATLAS IIAS/CENTAUR	Pad 36B	ICO-A1 (F2)
23 July 2001	ATLAS IIA/CENTAUR	Pad 36A	GOES-M
6 August 2001	TITAN IVB/IUS	Pad 40	DSP-21
11 October 2001	ATLAS IIAS/CENTAUR	Pad 36B	NRO MLV-12
16 January 2002	TITAN IVB/CENTAUR	Pad 40	MILSTAR (Flt 5)
21 February 2002	ATLAS IIIB/CENTAUR (1st ATLAS IIIB Launch)	Pad 36B	ECHOSTAR VII
8 March 2002	ATLAS IIA/CENTAUR	Pad 36A	TDRS-I
21 August 2002	ATLAS V/CENTAUR (1st ATLAS V Launch)	Pad 41	HOT BIRD 6)
18 September 02	ATLAS IIAS/CENTAUR	Pad 36A	HISPASAT-1D
5 December 2002	ATLAS IIA/CENTAUR	Pad 36A	TDRS-J
8 April 2003	TITAN IVB/CENTAUR	Pad 40	MILSTAR (Flt 6)
12 April 2003	ATLAS IIIB/CENTAUR	Pad 36B	ASIASAT-4
13 May 2003	ATLAS V/CENTAUR	Pad 41	HELLAS-SAT
17 July 2003	ATLAS V/CENTAUR	Pad 41	RAINBOW-1
9 September 2003	TITAN IVB/CENTAUR	Pad 40	NROL-19
18 December 2003	ATLAS IIIB/CENTAUR	Pad 36B	UHF F-11
5 February 2004	ATLAS IIAS/CENTAUR	Pad 36A	AMC-10
14 February 2004	TITAN IVB/IUS	Pad 40	DSP-22
13 March 2004	ATLAS IIIA/CENTAUR	Pad 36B	MBSat
16 April 2004	ATLAS IIAS/CENTAUR	Pad 36A	SUPERBIRD-6
19 May 2004	ATLAS IIAS/CENTAUR	Pad 36B	AMC-11
31 August 2004	ATLAS IIAS/CENTAUR	Pad 36A	NROL-1
17 December 2004	ATLAS V/CENTAUR	Pad 41	AMC-16
3 February 2005	ATLAS IIIB/CENTAUR	Pad 36B	NROL-23 (Final ATLAS III Flt)
11 March 2005	ATLAS V/CENTAUR	Pad 41	INMARSAT-4
30 April 2005	TITAN IVB/NUS	Pad 40	NROL-16 (Cape's Final TITAN Flt)
12 August 2005	ATLAS V/CENTAUR	Pad 41	MARS RECON ORBITER

19 January 2006	ATLAS V/CENTAUR	Pad 41	PLUTO NEW HORIZONS
20 April 2006	ATLAS V/CENTAUR	Pad 41	ASTRA 1KR

www.ingramcontent.com/pod-product-compliance
Lightning Source LLC
Chambersburg PA
CBHW080513110426
42742CB00017B/3104